POSTCRANIAL DESCRIPTIONS OF *ILARIA* AND *NGAPAKALDIA*
(VOMBATIFORMES, MARSUPIALIA) AND THE PHYLOGENY OF THE
VOMBATIFORMS BASED ON POSTCRANIAL MORPHOLOGY

Postcranial Descriptions of *Ilaria* and *Ngapakaldia* (Vombatiformes, Marsupialia) and the Phylogeny of the Vombatiforms Based on Postcranial Morphology

Carol J. Munson

UNIVERSITY OF CALIFORNIA PRESS
Berkeley • Los Angeles • Oxford

UNIVERSITY OF CALIFORNIA PUBLICATIONS IN ZOOLOGY

Editorial Board: Peter B. Moyle, James L. Patton,
Donald C. Potts, David S. Woodruff

Volume 125
Issue Date: January 1992

UNIVERSITY OF CALIFORNIA PRESS
BERKELEY AND LOS ANGELES, CALIFORNIA

UNIVERSITY OF CALIFORNIA PRESS, LTD.
OXFORD, ENGLAND

Library of Congress Cataloging-in-Publication Data

Munson, Carol J.
  Postcranial descriptions of Ilaria and Ngapakaldia (Vombatiformes,
Marsupialia) and the phylogeny of the vombatiforms based on
postcranial morphology / by Carol J. Munson.
       p.   cm. — (University of California publications in Zoology:
v. 125)
Includes bibliographical references.
ISBN 0-520-09772-6 (paper: alk. paper)
  1. Ilaria illumidens—Australia—Classification. 2. Ngapakaldia
tedfordi—Australia—Classification. 3. Paleontology—Miocene.
4. Paleontology—Australia. 5. Cladistic analysis. I. Title.
II. Series.
QE882.M3M86   1992   91-37565
569.2—dc20

Contents

List of Figures

List of Tables

Acknowledgments

This project began as a Master's thesis at the University of California at Riverside. Completion of the original thesis and preparations for publication could not have been successfully completed without the help of many people. I am indebted to Barbara Stein of the University of California Museum of Vertebrate Zoology at Berkeley, Lynn Barkley of the Los Angeles County Museum of Natural History, and Dave Kronen, osteologist, for the loans of specimens. I would like to thank Dr. Mike Greenwald, Dr. Howard Hutchison, and Bonnie Rauscher for their hospitality during my trips to Berkeley, as well as for lending me the fossils of *Ngapakaldia* and *Priscileo* for study, and I am grateful to Dr. Richard Tedford for the loan of *Ilaria* from the American Museum of Natural History. The librarians of the restricted book collection at UCLA were helpful in giving me access to references pre-dating 1930. Extensive editorial assistance was contributed by Rose Anne White and the editorial board of the University of California Press. I would especially like to thank Drs. Michael Woodburne, Rodolfo Ruibal, Dave Reznick, and Judd Case of the University of California at Riverside for their encouragement and advice; with extra thanks to Dr. Case for his constant guidance, support, and instruction, and to Dr. Woodburne for suggesting this project in the first place.

Abstract

The postcrania of the vombatiform marsupial *Ilaria illumidens* from medial Miocene strata of South Australia are described and compared to those of other vombatiforms, with the observation that *Ilaria* shares a similar morphology of the manus and pes with living wombats. While this indicates a certain degree of fossorial activity, the size and vertebral morphology of *Ilaria* argue against a burrowing lifestyle. Another medial Miocene vombatiform, *Ngapakaldia tedfordi*, is described as having a plesiomorphic vombatiform skeleton similar in many ways to that of phalangeriform possums, but with adaptations for greater size and a plantigrade, terrestrial habitus. Besides stouter and more robust limbs, these adaptations are evident in the concave dorsal surface and laterally facing fibular facet of the astragalus that creates a less flexible upper ankle joint.

For this study, a cladistic analysis was made using the postcrania of all the families in the Vombatiformes and several species representing outgroups, in order to establish synapomorphies uniting the group and to evaluate the position of these two genera within it. The results indicate that the ilariids and vombatids probably share a common ancestor, based on the similarity of the metapodials and phalanges, especially the uniquely identical morphology of the proximal metapodial facets. *Ngapakaldia*'s similarity in form to phalangeriform possums reflects the arboreal ancestry of the vombatiform clade and indicates the plesiomorphic state from which the postcrania of other, more specialized vombatiform families (i.e., fossorial wombats and ilariids) are derived.

INTRODUCTION

The primary purpose of this study is to describe the postcranial anatomy of the vombatiform marsupial *Ilaria illumidens* (Tedford and Woodburne 1987) from medial Miocene strata of South Australia. To establish a frame of reference, a comparative study of the postcrania of all the vombatiform families was made. This resulted not only in a better understanding of the phyletic position of *Ilaria*, but also in recognition of a number of synapomorphies for the Vombatiformes as a whole and for the several families which comprise it. The postcranial elements of the medial Miocene palorchestid *Ngapakaldia tedfordi* (Stirton 1967) have never been fully described, even though the species has been well known for some time. As the plesiomorphic vombatiform features of the postcrania of *N. tedfordi* provide a directional contrast to the apomorphic features of *Ilaria*, a formal description of *N. tedfordi* was undertaken as well.

The original descriptions of both of these species and current phyletic studies of the vombatiforms are based on a combination of cranial and dental characters. While there is no question of the value of postcranial material in reconstructing the appearance and habitus of an animal, data developed during the course of this study, as well as those of other workers (Gregory 1910, 1951; Szalay 1977, 1981, 1982, 1984; and Lewis 1964, 1980a,b, 1983), indicate that postcranial characters, especially those of the manus and pes, have great phylogenetic significance.

Like teeth, podials are complex—that is, they demonstrate a number of diverse characters that have changed over time in response to selective pressures. These characters can be analyzed not only for synapomorphies based on outgroup comparisons, but more important, in conjunction with the stratigraphic position of the specimens, they can also be analyzed for transformational sequences (evolutionary patterns that help establish the most likely phylogenetic hypotheses; Szalay 1984). Because podials, and to a lesser extent metapodials, are compact, they are fairly resistant to damage and therefore are frequently found in fossil assemblages (Behrensmeyer and Boaz 1980).

Table 1. Classification of the Diprodontia according to
Woodburne (1984) and Marshall, Case, and Woodburne (1990)

Suborder Phalangeriforms: possums, gliders, kangaroos
Suborder Vombatiformes
     Superfamily Phascolarctoidea
          Family Phascolarctidae: koalas (arboreal)
     Superfamily Vombatoidea
          Family Thylacoleonidae (digitigrade,carnivorous)
          Family Ilariidae
          Family Wynyardiidae
          Family Vombatidae: wombats (fossorial)
          Family Palorchestidae (plantigrade, lophodont)
          Family Diprotodontidae (graviportal, lophodont)

Among the members of the Vombatiformes, it can be demonstrated that
elements of both the manus and the pes have characters useful in establishing
phylogenetic relationships. Each of the seven families in this group (described
below) is characterized by distinct locomotor adaptations, including arboreal,
fossorial, digitigrade, plantigrade, and graviportal morphologies. In addition to the
unique character complexes that can be used to identify each family, the manus and
pes also reflect synapomorphies both between families and with the sister-group of
the Vombatiformes, the Phalangeriformes. Comparison of the entire postcranium,
but especially the feet, between both living and extinct taxa can establish polarities
and developmental trends within the suborder.

CLASSIFICATION
(table   1)

The Vombatiformes and the Phalangeriformes are the two suborders of the
marsupial Order Diprotodontia (classification after Woodburne 1984). The order is
characterized, osteologically, by two prominent lower incisors and a syndactylous
condition of the pes (reduced second and third digits enclosed within a common
sheath of skin). The Suborder Phalangeriformes includes the modern possums,
phalangers, gliders, and kangaroos. The Suborder Vombatiformes consists of two
superfamilies: the Phascolarctoidea (koalas) and the Vombatoidea (wombats and
their fossil relatives).

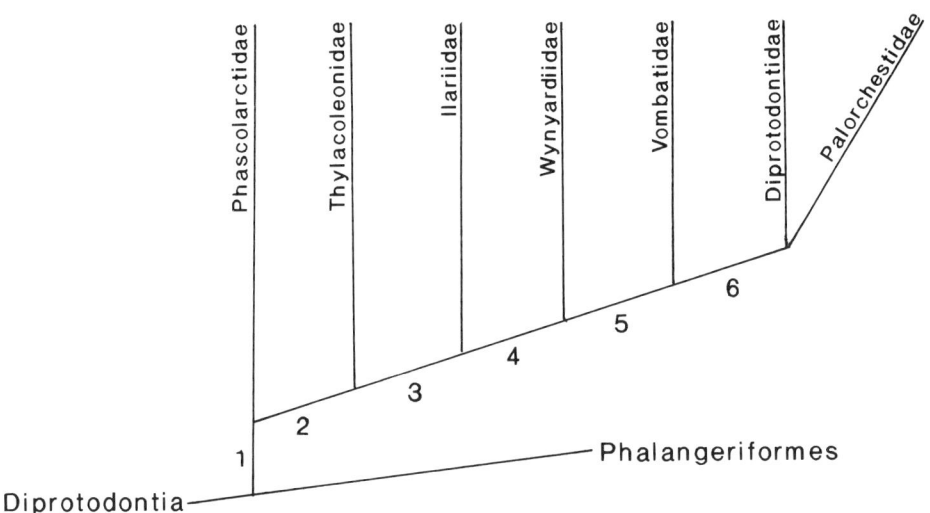

Fig. 1. Cranial apomorphies of the Vombatiformes according to Marshall, Case, and Woodburne (1990). Numbers correspond to nodes on the cladogram: 1. Palatal vacuities enclosed by palatine bones; dentary shortened between I/1 and P/3; I/1 steeply inclined; reduction of functional premolars to P3 above and below; functional dental formula of I3/1, C1/0, P1/1, M4/4. 2. Bilaminar bulla, squamosal extends into roof of alisphenoid hypotympanic sinus. 3. Squamosal bulla; pterygoid flange of dentary expanded; loss of fine crenulations from molar crowns; loss of P1, P/2. 4. Lowering of inclination angle of I/1; reduction of paralophid shear; trend toward lower molar bilophodonty; loss of P2/. 5. Hypertrophy (elongation) and increased ankyloses of dentary symphysis; diastema elongate; loss of C1/. 6. Palatal vacuities lost; molars completely lopohodont.

The Phascolarctoidea consists of one family, the Phascolarctidae, with postcranial material being known from only one species, the modern koala, *Phascolarctos cinereus*. Although the limbs of the koala resemble those of the phalangers (they share the consequences of the same arboreal habitus), certain similarities between koala and wombat anatomy have long been recognized (Owen 1838, Forbes 1881, Sonntag 1923, Strahan 1976, Bryant 1977, Barbour 1977). Recent work in serology (Kirsch 1977) and spermatozoan structure (Hughes 1965; Harding, Carrick, and Shorey 1979) have confirmed the koalas' close relationship to the wombats.

The Vombatoidea contains six families: the Thylacoleonidae (carnivorous "marsupial lions"), the Ilariidae, the Wynyardiidae, the Vombatidae (wombats), the Palorchestidae, and the Diprotodontidae. The vombatoids are united by characteristics of their cranial anatomy, particularly the contribution of the squamosal bone to all or part of the auditory bulla.

Within the Vombatoidea, the relationships among families are not clear. A recent phylogenetic scheme by Marshall, Case, and Woodburne (1990) suggests grouping the five herbivorous families (all but the thylacoleonids) on the degree of lophodonty expressed in the molars (see fig. 1). Accepting the proposition that selenodont teeth, as seen in the phascolarctids, are plesiomorphic for the suborder (Archer 1984 a,b), a transformational sequence of increasing lophodonty is seen to progress from the ilariids through the wynyardiids to the fully lophodont palorchestids and diprotodontids. The highly specialized rodent-like shape of the vombatid cheek teeth has masked their position on this transformational sequence, but they are classified with the palorchestids and diprotodontids, based on similarities of jaw structure.

Except for noticing general resemblances between the limb bones and tarsals of vombatids, wynyardiids, palorchestids, and thylacoleonids (Finch 1982, Archer 1984a, Pledge 1987), the postcranial skeleton has not been used before to evaluate phylogenetic relationships within the Vombatiformes.

TERMINOLOGY AND MATERIALS

TERMINOLOGY

Anatomical terms are from Sisson (1914), Gray (1977), and Spence and Mason (1983), with the following exceptions. In the pes, the terms entocuneiform, mesocuneiform, and ectocuneiform are used for the medial, intermediate, and lateral cuneiforms, respectively. In diprotodontian marsupials, the scaphoid and lunate bones of the manus are fused into a single scapholunar.

Definitions of anatomical directions are depicted in Fig. 2. Length is always measured on the long axis of the axial skeleton and the long axis of the limbs (including the podials, where it is not always the longest dimension). Width is always measured mediolaterally, and height (or thickness) is in the third plane, whether it be dorsoventral (axial skeleton) or anteroposterior (limbs).

There seems to be little agreement between authors on the use of the terms dorsal/ventral and anterior/posterior in regard to quadruped feet. This study will assume that the animal is in a natural, four-footed plantigrade posture, so that the "back of the hand" (or foot) is dorsal, and the plantar surface is ventral. The same terminology will apply to the podials as to the metapodials. To avoid confusion, the unambiguous terms proximal and distal will be used whenever possible, in place of anterior and posterior.

MATERIALS

A list of all specimens observed is given in Appendix I. An attempt was made to compare as many vombatiform specimens as possible, both male and female, juvenile and adult. A number of other australidelphian marsupials were also studied for their value as outgroups. The number of fossil specimens observed is low because of the scarcity of positively identified postcranial remains. In some cases, descriptions from other studies have been used to supplement or to replace actual specimens (i.e., *Phascolonus*, *Diprotodon*).

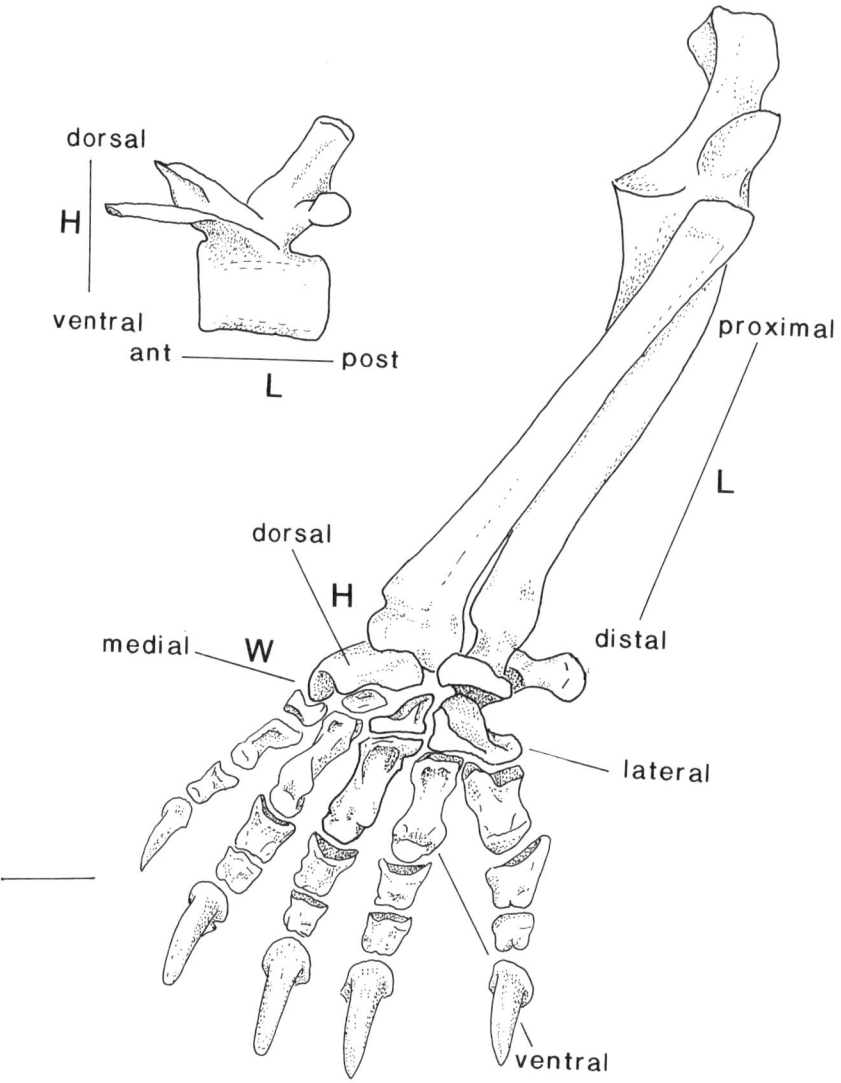

Fig. 2. Definitions of anatomical directions used herein. The left forearm and manus and 4th lumbar vertebra of *Vombatus ursinus*, the common wombat. Scale = 2 cm. Abbreviations: ant, anterior; H, height (or thickness); L, length; post, posterior; W, width.

NGAPAKALDIA
Description and Comparisons

Stirton (1967) described the medial Miocene palorchestids *Ngapakaldia tedfordi, N. bonythoni,* and *Pitikantia dailyi* collected from various sites in the Etadunna Formation around Lake Ngapakaldi and Lake Pitikanta in the Lake Eyre Basin, South Australia. They are part of the Ngapakaldi Fauna of medial Miocene age (ca. 14-17 Ma; Woodburne et al. 1985). The smallest of the three, the sheep-size *N. tedfordi,* left numerous postcranial remains preserved, including complete feet still in the articulated position (Rich et al. 1985). There were no complete skeletons, but different elements from several individuals provide a very good picture of the postcranium of the species. The left radius, ulna, complete manus, and nearly complete left pes were associated with the holotype cranium, SAM P13851 (study was done from casts of postcranial material, UCMP 126722). A nearly complete left pes, questionably referred to the slightly larger taxon *P. dailyi,* was recovered, as were the forearm, manus, and pes of the even larger, calf-size *N. bonythoni.*

Waters (1967) compared *N. tedfordi* to *Diprotodon optatum* and other marsupials, concluding that the former most resembled *Phascolarctos cinereus* and *Didelphis virginianis.* This is true only to a very limited degree, and acknowledges that all three share some generalized plesiomorphic features of all marsupials. No phalangeriforms were used in Waters'study, and thus the striking similarity of *N. tedfordi*'s postcrania to the more primitive members of this suborder was not noticed. In particular, *N. tedfordi* resembles the semi-terrestrial common brush-tailed possum *Trichosurus vulpecula,* especially in elements of the manus and pes. *Ngapakaldia*'s relationship to the vombatiforms is evident in the adaptations of the postcranial skeleton for increased size and complete terrestriality imposed upon this generalized phalangeriform pattern.

VERTEBRAL COLUMN

Few vertebrae of *Ngapakaldia* are available. There is a partial cervical series (UCMP 69812), two thoracic vertebrae and a left rib (UCMP 71416), one lumbar vertebra (UCMP 72130), a fragment of the sacrum (UCMP 69813), and a few caudal vertebrae (UCMP 72130 and 69813).

Cervical vertebrae (fig. 3)

The cervical vertebrae include fragments of the atlas; nearly complete axis, ?third, and ?fourth cervicals; and fragments of at least two others. The atlas fragments are from the left side and consist of the left posterior articular facet, part of the left anterior facet, a portion of the pedicle and lamina between the two, and the transverse process. The ridge between the left anterior and posterior articular facets is low and smooth, indicating that the atlas was not anteroposteriorly shortened as in wombats. The transverse process curls up slightly at the tips and is relatively thin dorsoventrally.

The axis (fig. 3A) is missing only the spinous and transverse processes. The width of its ventral surface is approximately equal to the length without the odontoid process. This is the same as in *Macrotis lagotis* (the greater bilby), *Trichosurus*, *Phascolarctos*, and *Diprotodon*. In comparison, the width is about .50 greater than the length (short neck) in wombats, *Thylacoleo*, *Pseudocheirus* sp. (ring-tailed possum), and *Wynyardia*.

The ?third cervical (fig. 3B) is quite complete, lacking only the tip of the left transverse process. On the right side, the process is complete, enclosing a transverse foramen. The overall form of the cervical vertebrae available is identical with that in *Trichosurus vulpecula*, the common brush-tailed possum. However, there is a difference in size, as the linear measurements of *Ngapakaldia* are two and a half to three times greater than those of *Trichosurus*. When articulated and viewed laterally, the cervical vertebrae of both *Trichosurus* and *Ngapakaldia* show gaps between the spinous processes of successive vertebrae, indicating a similar relative length of the neck. In short-necked animals, like wombats and ring-tailed possums, the vertebrae are closely pressed to one another, with little or no space between the neural spines.

Thoracic vertebrae (fig. 3)

There are two thoracic vertebrae, estimated by comparison with the brush-tailed possum to be vertebrae one and six of the series. A full vertebral column was unavailable for this study, but Waters (1967) reports that *N. tedfordi* has thirteen

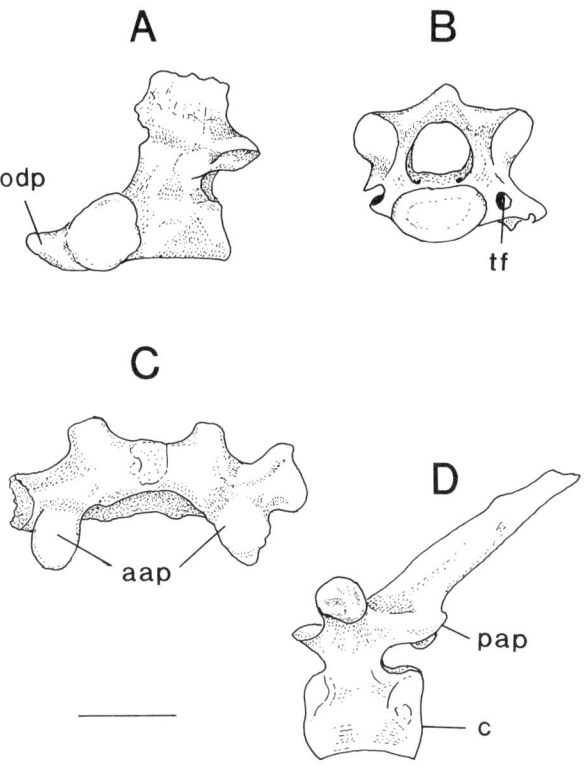

Fig. 3. *Ngapakaldia. tedfordi*: A and B. cervical vertebrae, UCMP 69812; C and D. thoracic vertebrae, UCMP 71416. A. axis, left lateral view; B. third cervical vertebra, posterior view; C. first thoracic vertebra, dorsal view; D. sixth thoracic vertebra, left lateral view. Scale = 2 cm. Abbreviations: aap, anterior articular processes; c, centrum; odp, odontoid process; pap, posterior articular process; tf, transverse foramen.

thoracic vertebrae—which appears to be the plesiomorphic number for all marsupials. The ?first thoracic vertebra (fig. 3C) is somewhat eroded and has lost the spinous process and the posterior half of the centrum. It is typical of first thoracic vertebrae for diprotodontian marsupials in that the anterior articular facets are spaced very wide apart, the transverse processes are round and bulbous, and the centrum is much wider than it is high. The centrum of the ?sixth vertebra (fig. 3D) is much deeper, similar in transverse section (width/height ≈ .80) to those of *Thylacoleo* and *Ilaria* but longer anteroposteriorly, like *Trichosurus*. It has well developed demi-facets for rib heads, both anterior and posterior, and complete articular, transverse, and spinous processes. Viewed dorsally, the posterior articular processes do not diverge widely from the neural spine as they do in *Trichosurus* and *Phascolarctos*; rather, they are closer and less prominent, a feature common to all the vombatoids observed.

Lumbar vertebrae (fig. 4)

The lumbar vertebra is complete except for the right transverse process and the edges of the spinous process. The centrum is dorsoventrally compressed, wider than it is high, though not so much as in wombats. Length of the centrum is the greater dimension, as it is in *Trichosurus*, wombats, and *Phascolarctos*. The short, curving shape of the transverse process is the same as in *Trichosurus*, but instead of curving forward and down the process remains in the frontal plane, as it does in all the other vombatiforms.

The position of this vertebra is uncertain, but Waters (1967) reports a total of six lumbar vertebrae for *Ngapakaldia*, which with the thirteen thoracics is consistent with the total of nineteen thoracolumbar vertebrae seen in all marsupials (Dublin 1903).

Sacral vertebrae

The sacrum is missing most of the first vertebra, including the wings. The ventral half of the second vertebra is connected to a nearly complete third sacral vertebra, forming one complete sacral foramen on the right side. It is impossible to tell if there was a fourth sacral vertebra as in an adult *Phascolarctos*, but from the shape of the third it appears unlikely. Three fused vertebrae (only two of which articulate with the ilium) is the most common number in all the species studied. The centra are long and dorsoventrally compressed, most similar in shape to those of *Trichosurus*.

Caudal vertebrae

There are four caudal vertebrae (UCMP 69813) attached to a sacrum, and two from another specimen (UCMP 72130) that appear to be the first two of the series. They all have broad transverse processes that project almost at right angles to the centra, and then curve slightly posteriorly. The neural canals of all are complete and circular. The centra of the first two caudal vertebrae are flattened dorso-ventrally, wider than high in transverse section. The third and fourth caudals are deeper, more circular in cross-section. All of the centra are longer than they are wide, the

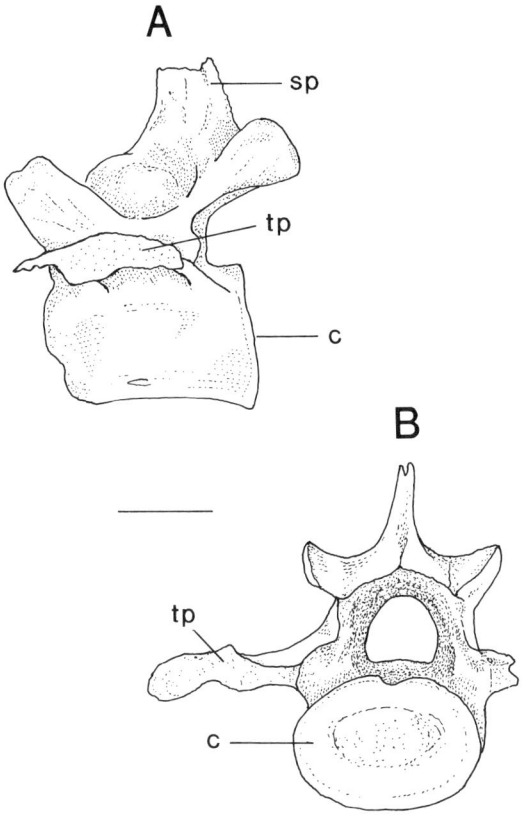

Fig. 4. *N. tedfordi*, lumbar vertebra, UCMP 72130: A. left lateral view; B. posterior view. Scale = 2 cm. Abbreviations: c, centrum; sp, spinous process; tp, transverse process.

third and fourth being particularly spool-shaped. There is a faint indication of a ventral groove on the centra, most obvious on the fourth caudal (also mentioned by Waters 1967), that indicates the presence of chevron bones in life. In all characteristics, the caudal vertebrae of *Ngapakaldia* are practically identical with those of *Trichosurus*, and quite unlike those of *Phascolarctos* or the wombats. The number of caudal vertebrae is also similar, as *Trichosurus* has 29 and Waters (1967) reports ± 26 for *Ngapakaldia*. Wombats, of course, have shortened tails with only 11-13 caudal vertebrae; koalas' are even shorter, with only 5-8.

In summary, the vertebral column of *Ngapakaldia* largely maintains the plesiomorphic structure of the arboreal phalangeriforms. There are only two derived character states in this series of bones: the position of the transverse processes in the lumbar vertebrae (see table 2), and the nondivergent posterior articular processes of the thoracic vertebrae (table 3).

STERNUM

As represented by the material at hand, the sternum of *Ngapakaldia* (UCMP 71416) is incomplete. Only the manubrium, broken in the middle, and one fragment of a body element are present. Because of erosion at the ends of the piece presumed to be the posterior half of the manubrium, there is the slight possibility that it could be another body element instead. If indeed it does represent the remainder of the manubrium (and the fit is good, if not exact), then the dimensions and shape of the manubrium would again be identical with those of *Trichosurus*. If, however, the manubrium is incomplete, then it would be impossible to determine whether it is short and broad as in *Phascolarctos*, the wombats, and *Diprotodon*, or long and slender as in the other species studied. The confirmed body element is triangular in transverse section, as is the manubrium, indicating that it is probably the most anterior element of this section.

FORELIMB

Though the axial postcranial skeleton was most like that of *Trichosurus*, only larger, the appendicular skeleton begins to show some interesting differences. The overall appearance is still very phalangeriform-like, but functional adaptations for terrestriality and greater strength of the forelimb are also evident.

Humerus (fig. 5)

The humerus is a good example of these adaptations. There are two humeri, the right and left, from the same animal (UCMP 71416). The right one is most complete, with an intact distal end and a long deltoid ridge, but both tuberosities of the proximal end are missing. The left humerus has a lesser tuberosity, but no greater, a broken deltoid ridge and a fragmented distal end. The general shape of the deltoid ridge, and distal condyles is most like that in *Trichosurus* and *Phascolarctos*. While its relative length is comparable to that in wombats and bandicoots, the deltoid ridge does not have the lateral overhang seen in wombats or *Diprotodon*, and the trochlea is not greatly expanded onto the posterior side as in wombats.

The width across the epicondyles at the distal end, however, is much broader than in phalangers, bandicoots, or dasyurids. Of the species studied, only in the

Table 2. Apomorphies of the Suborder Vombatiformes

		Vombatiformes	Plesiomorphic state
1.	Transverse processes of lumbar vertebrae	Perpendicular to sagittal plane	In sagittal plane
2.	Width, distal end of humerus	> 0.3 length	< 0.3 length
3.	Position of the cuneiform (manus)	Higher (more proximal) than scapholunar	Directly lateral to scapholunar
*4.	Width of MC I	> 0.5 length	< 0.5 length
*5.	Length of tibia	< length femur	≥ femur
*6.	Distal surface of tibia (astragalar facet)	Ligamental pit posterior to medial malleolus	Surface completely smooth
7.	Orientation of tibial crest	Between digits II and III	Between I and II
*8.	Fibular facet of astragalus	Reduced and lateral	Large and on dorsal surface
*9.	Lateral tibial facet of astragalus	Concave	Convex
10.	Entocuneiform facet of MT I	Strong concave component	Mostly convex
11.	Length of proximal phalanges	Much shorter than corresponding MC or MT	Nearly as long as or longer than corresponding MC or MT
12.	Length of medial phalanges	< 0.5 length MT III	> 0.5 length MT III

(*) indicates weighted characters in the phylogenetic analysis

Table 3. Apomorphies of the Superfamily Vombatoidea

	Vombatoidea	Plesiomorphic state
1. Posterior articular processes of anterior thoracic vertebrae	Close to midline (viewed dorsally)	Diverge from midline
2. Coracoid process of scapula	Reduced	Extends medial to glenoid fossa
3. Size of triceps notch, scapula	Increased, large	Small
*4. Length of radius	≤ humerus	> humerus
5. Thickness (ant-post) of radius (distal end)	≥ 0.1 length	< 0.1 length
*6. Thickness (ant-post) of ulna at semilunar notch	> 0.1 length of ulna	< 0.1 length ulna
*7. Width (med-lat) of pisiform	> 0.5 ant-post thickness	< 0.5 thickness
*8. Area, MC V facet of unciform	≥ MC IV facet	< MC IV facet
*9. Width, distal end of MC V	≥ 0.5 length	<< 0.5 length
10. Width of femur, proximal end	≥ 0.3 length	< 0.3 length
11. Lateral notch, distal end of fibula (for extensor tendons)	Reduced (except in Wynyardia bassiana)	Deep
12. Area, MT V facet of cuboid	≥ MT IV facet (except Lasiorhinus)	< MT IV facet

(*) indicates weighted characters in the phylogenetic analysis

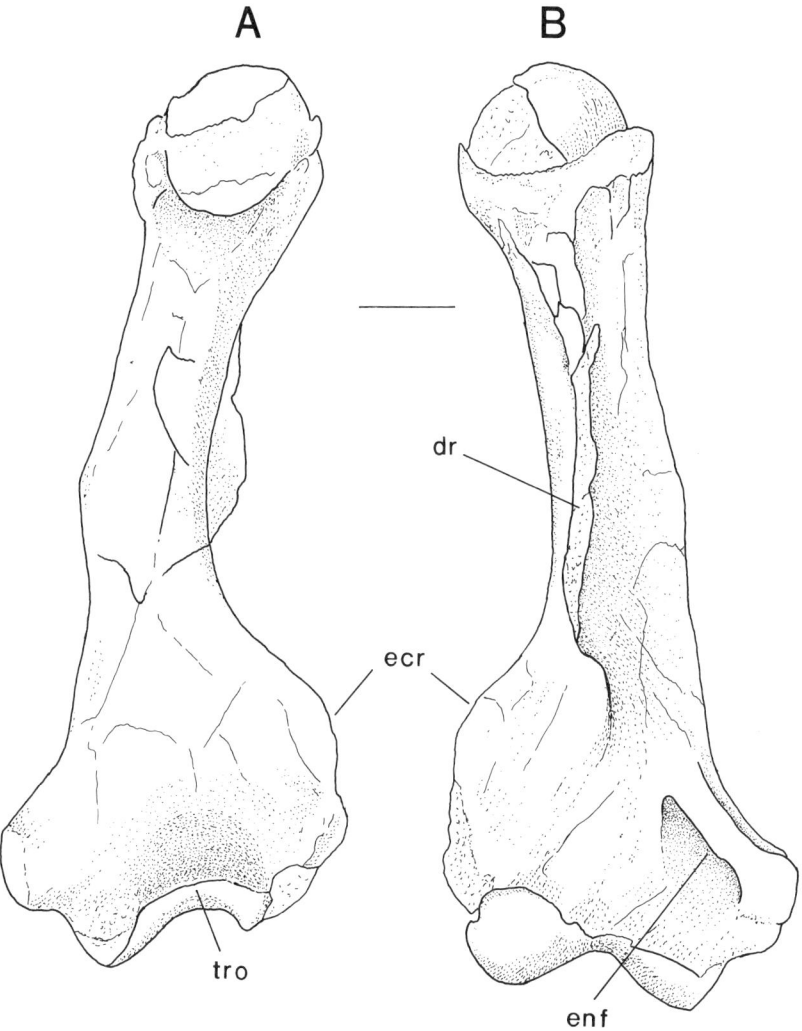

Fig. 5. *N. tedfordi*, right humerus, UCMP 71416: A. posterior view; B. anterior view. Scale = 2 cm. Abbreviations: dr, deltoid ridge; enf, entepicondylar foramen; ecr, ectepicondylar ridge; tro, trochlea.

vombatiform humeri is there a maximum distal width-to-length ratio greater than 0.3, and in *Ngapakaldia*, the wombats, and *Priscileo*, the ratio is highest at 0.4-0.46. Broad epicondyles provide greater attachment area for the muscles of the forearm, indicating greater strength for flexion, extension, pronation, and supination of the manus (Hildebrand 1988).

Ulna (fig. 6A, B)

The ulna of *Ngapakaldia* is much stouter than in the arboreal phalangeriform species. There are two ulnae: one lacking the distal end (UCMP 71416) and one lacking the proximal end (UCMP 126722). The shaft has a medially concave curvature as in wombats, *Ilaria*, and *Diprotodon*. The anteroposterior thickness of the shaft at the semilunar notch is the same as in all the vombatoids (i.e., greater than 0.1 the length of the shaft). The length of the olecranon process is approximately 0.2 the length of the shaft, the same as in *Ilaria*, compared to only 0.1 in the phalangers and the small dasyurids studied (in the wombats it equals 0.25). Unlike wombats, the anconaeus process is small and the radial facet flat. These features, combined with the smaller trochlea of the distal humerus, would appear to indicate that much less force is being applied at the elbow joint than in wombats.

Radius (fig. 6C)

The radius (UCMP 126722) is complete, its structure primitive for vombatiforms and similar to the arboreal phalangeriforms. The shaft is nearly straight, as in *Trichosurus*, but with a slight lateral curve, more like that seen in *Ilaria*. The distal end is only slightly thicker (anteroposteriorly) and considerably broader (mediolaterally) than in *Trichosurus*, but not as much in either dimension as in the wombats. In these features, it is more similar to *Ilaria* and *Thylacoleo*.

In order to estimate the ratio of the length of the upper arm to the forearm, it was necessary to use data from two different individuals. One (UCMP 71416) has a complete humerus and an incomplete ulna, the other (UCMP 126722) has another partial ulna and a complete radius. Since the two ulnae appear to be comparable in size, a comparison can be made between the right humerus (UCMP 71416) and the left radius (UCMP 126722). The two are nearly equal in length; thus the estimated radial/humeral length ratio is 0.94, which is quite close to the measurement in the wombats (0.94-1.02) and the two thylacoleonids (1.04).

In the arboreal species, the radius is longer than the humerus (rl/hl averages 1.12 in the three *Trichosurus* and five *Phascolarctos* specimens, 1.3 in the two *Phalanger sp.* specimens), except in *Pseudochirus* (rl/hl = 1.03; only one specimen studied). A statistical comparison of radial/humeral length between the vombatoids and two

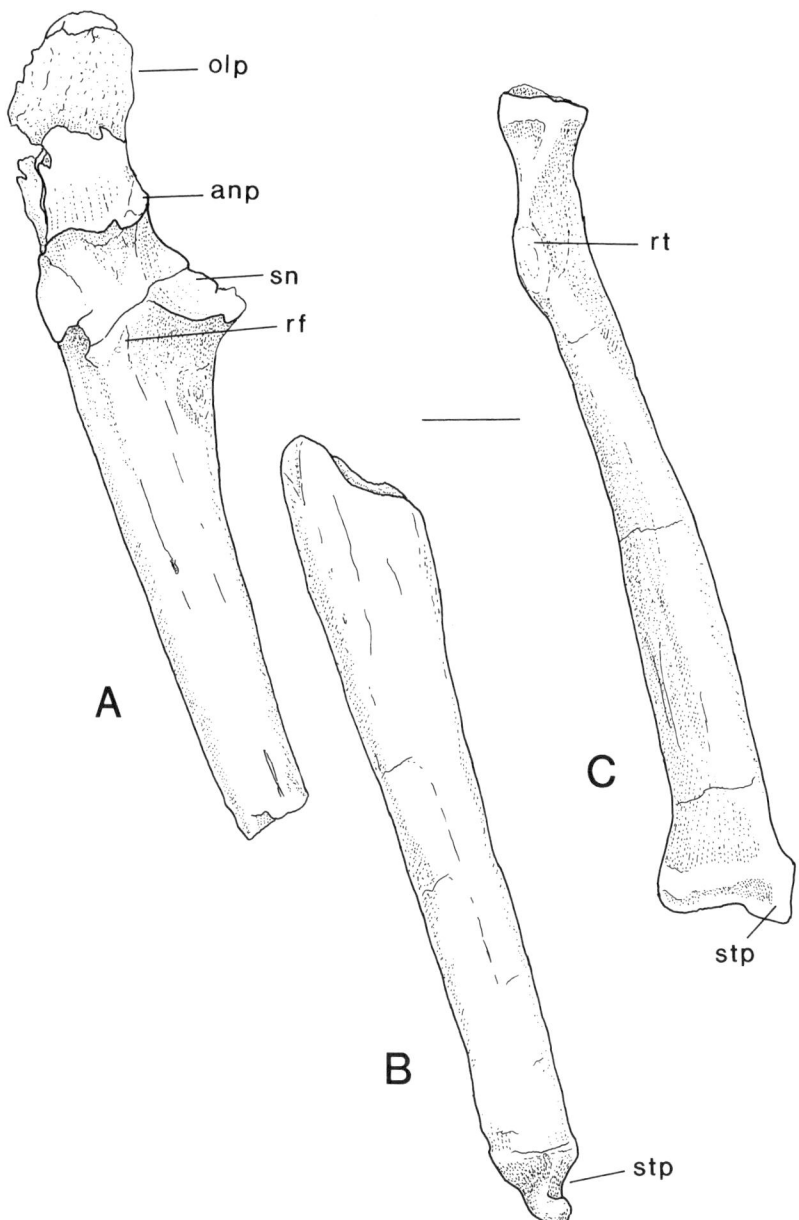

Fig. 6. *N. tedfordi*: A. right proximal ulna, anterior view, UCMP 71416; B. left distal ulna, anterior view; C. left radius, posterior view; UCMP 126722. Scale = 2 cm. Abbreviations: anp, anconaeus process; olp, olecranon process; rf, radial foramen; rt, radial tuberosity; sn, semilunar notch; stp, styloid process.

closest outgroups (phascolarctids and phalangeriforms) shows a significant difference (T-test = -2.32; probability = 0.025). The mean ratio of radial/humeral length
in the outgroup is 1.1; in the vombatoids (wombats and thylacoleonids) the mean
equals 0.94. The difference between the two group means is even greater if
measurements for the graviportal *Diprotodon* (rl/hl = 0.64-0.67) are taken into
account, as these data would decrease the mean for the vombatoids.

A shorter radius increases the strength of the forelimb by decreasing the length of
the out-arm at the elbow joint, and appears to be a synapomorphy of the vombatoids
(see table 3). The increased length of the olecranon process (over that of the
arboreal species) increases the in-arm at this same joint, also contributing to greater
mechanical advantage of the forelimb.

In summary, the morphology of the forelimb shows certain terrestrial derivations
that reach an extreme state in the wombats and *Diprotodon*. The shape of the
humerus is still most similar to that in the arboreal phalangeriforms, except for the
broad distal epicondyles (table 2). The ulna and radius, however, are more similar
to those of the terrestrial vombatoids, and show a number of shared derived
characters: the increased thickness of the ulna at the semilunar notch, the
decreased relative length of the radius, and the increased thickness of the distal end
of the radius (table 3). The increased length of the olecranon process, while not
shared by the graviportal *Diprotodon*, may still be a synapomorphy for the
palorchestids, ilariids, and vombatids (table 4).

MANUS
(fig. 7)

There is a complete left manus (UCMP 126722) from the same individual (the holotype) as the radius and distal ulnar element. All observations were made from casts
of the original bones, which appear to have been in excellent condition. In reconstructing the manus of *N. tedfordi*, the pattern of articulation seen in *Trichosurus* (fig.
8) was used, after noticing the nearly identical shape (except for size) of the
individual carpals and the proximal facets of the metacarpals. The few differences
that do exist between the forefeet of *Ngapakaldia* and *Trichosurus* are important,
because in each case they can be interpreted as adaptations for a completely
terrestrial habitus and increased size.

Carpals

Of the carpals, the greatest differences between the above two species are seen in
the pisiform and cuneiform bones. In *Ngapakaldia*, the ulnar facet of the pisiform
is a slightly concave, mediolaterally wide triangle which when articulated with the

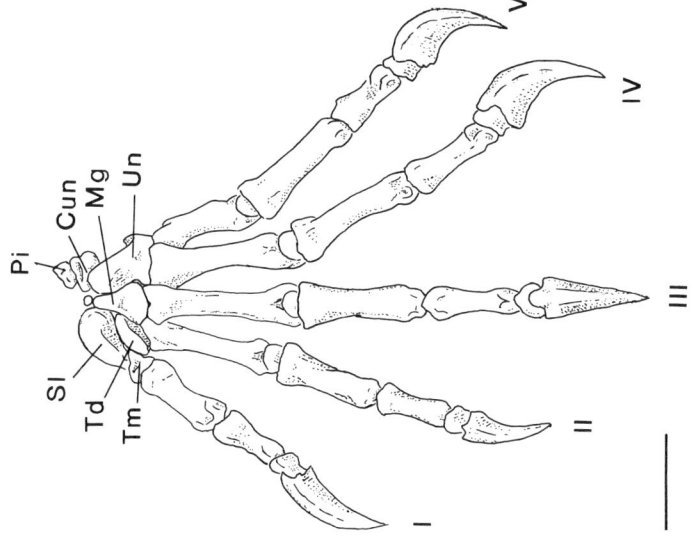

Fig. 8. *Trichosurus vulpecula*, left manus, MVZ 127142. Scale = 1 cm. Abbreviations as in fig. 7.

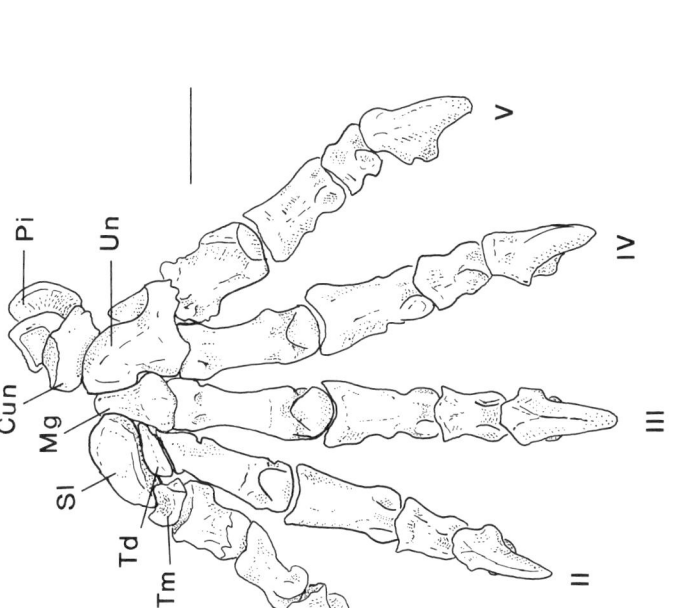

Fig. 7. *N. tedfordi*, left manus, UCMP 126722, dorsal view. Note the greater similarity of the carpals and the ends of the metacarpals to those of *Trichosurus* (fig. 8) than to those of *Vombatus* (fig. 2). Scale = 2 cm. Abbreviations: Cun, cuneiform; Mg, magnum; Pi, pisiform; Sl, scapholunar; Td, trapezoid; Tm, trapezium; Un, unciform.

Table 4. Apomorphies of the Palorchestids, Diprotodontids, Ilariids, and Vombatids

	Palorchestids, Diprotodontids, Ilariids and Vombatids	Plesiomorphic state
1. Length, olecranon process of ulna	≥ 0.2 length ulna (except *Diprotodon*)	< 0.2 length ulna
2. Size, ulnar facet of pisiform	≥ size cuneiform facet	< cuneiform
*3. Metacarpals	Relatively short and stout, rugose, length of MC III < 3X mid-shaft width	Long and slender, length MC III > 3X width
4. Facet for MC IV on MC III	Dorsal to ligamental pit	Proximal to ligamental pit
5. Length of MC IV and digital formula (longest to shortest) of manus	Reduced 3.2.4.5.1 (except *Diprotodon*)	3.4.2.5.1 phalangers 3.4.5.2.1 *Phascolarctos*
*6. Width of iliac crest	≥ 0.5 length of ilium	< 0.5 length of ilium
*7. Width, proximal end of fibula	≥ 0.1 length of fibula	< 0.1 length of fibula

Table 4 (cont.)

*8.	Angle between femoral facet and fibular sesamoid facet on fibula	> 100° (also in *Wynyardia*)	< 100°
9.	Length (prox-dist) of ectocuneiform	Reduced < height (dorsoventral)	≥ height
10.	Maximum width of MT V	≥ 0.7 length	< 0.5 length
*11.	Length of MT V and digital formula of pes	Reduced 4.3.2.5.1 (except *Diprotodon*)	4.5.3.2.1 phalangers 4.3.5.2.1 *Phascolarctos*
12.	Mid-shaft width of proximal phalanges (IV and V only in pes)	≥ 0.4 length of shaft	< 0.4 length of phalanx

(*) indicates weighted characters in the phylogenetic analysis

cuneiform helps to form a shallow socket for the styloid process of the ulna. In *Trichosurus*, this facet is relatively narrow and rectangular and appears to have little or no articulation with the ulna. This articulation is also weak and the pisiform narrow (width < 0.5 length) in *Phalanger, Phascolarctos, Macrotis,* and *Dromiciops,* animals whose size or lifestyle causes them to put little weight on the front feet. In all the vombatoids (animals which tend to be large and are fully terrestrial), the pisiform is wide (width > 0.5 length) and the ulnar articulation strong, as seen in *Ngapakaldia.*

The cuneiform of *Ngapakaldia* is a thick triangular bone, the proximal surface divided by a biconcave ridge into two nearly equal concave facets for the ulna and pisiform. When the carpus is articulated, the cuneiform sits higher (more prox-imally) than the scapholunar, resulting in a broader contact of the cuneiform with the radius than is seen in any of the outgroups studied. In this aspect, and in general outline, the cuneiform appears most similar to other vombatiforms; however, it is thick proximodistally, not flattened as in wombats and *Thylacoleo.*

The unciforms of *Ngapakaldia* and *Trichosurus* are very similar in shape: both are approximately as long (proximodistally) as they are wide (mediolaterally), with a large hamate on the posterolateral corner. The only difference is that the meta-carpal V facet is equal in size to the MC IV facet in *Ngapakaldia,* whereas the MC IV facet is the larger of the two in *Trichosurus.* This reflects the increased size of digit V common to all the vombatoids.

The scapholunar, trapezium, trapezoid, and magnum of *Ngapakaldia* are larger, but identical in morphology with those of *Trichosurus.* The scapholunar is teardrop-shaped: rounded on the lateral end, tapered medially. The medial tip fits into a concave posteromedial notch on the trapezium, and a flat distal facet articulates with the trapezoid.

The trapezium is roughly triangular, with curving saddle-shaped facets for MC I distally and the trapezoid laterally. When articulated, most of the proximal surface is exposed, as the scapholunar sits across the posterior corner only. The antero-dorsal tip just barely touches MC II. There probably was a ligamentous attachment here as there is in *Trichosurus.*

The trapezoid is a biconvex, lenticular, triangular bone. It sits at an angle in the manus because of the mediolaterally tapered proximal end of MC II. Its thicker medial side rests on the trapezium, the proximodistally tapered lateral side rests on the base of the magnum. The posterior half of the proximal surface fits under the edge of the scapholunar.

The magnum is longer proximodistally than it is wide—unlike the wombats, in which it is wider than long (proximodistally flattened). The distal surface of the magnum has a triangular saddle-shaped facet for MC III, with a faint ridge or keel running dorsoventrally. A mediolaterally compressed projection rises proximally from the base. It articulates to the unciform laterally by a saddle-shaped facet; the convex medial surface is covered by the scapholunar. The posterior third of both the medial and lateral sides is a rugose area for ligament attachment.

Metacarpals (figs. 9C and 10D)

The shapes of the proximal and distal ends of *Ngapakaldia*'s metacarpals, especially II, III, and IV, are nearly identical with those in *Trichosurus* (figs. 9D and 10C). The proximal end of MC II in both genera is mediolaterally tapered, so that the trapezoid facet faces proximomedially, and laterally there is broad contact with MC III and only minor contact with the magnum. The proximal end of MC III in both species and in the microbiotheriid *Dromiciops* has a concave border in dorsal aspect and a saddle-shaped triangular magnum facet. There is only minor contact with the unciform. The unciform facet of MC IV in all three genera is also saddle-shaped with a concave dorsal border. By comparison, the proximal facets of MCs III and IV in wombats and *Ilaria* are smoothly convex, MC II is not tapered, and it has a large facet for contact with the magnum (see figs. 9 and 10 and table 5). The distal condyles of all the metacarpals in *Ngapakaldia*, *Trichosurus*, and *Dromiciops* are high dorsoventrally and spherical. This is in marked contrast to the dorsoventrally flattened condyles of the wombats, *Ilaria*, *Phascolarctos*, and *Diprotodon* (table 5).

Imposed upon this probably plesiomorphic pattern, *Ngapakaldia* nevertheless shows some distinctive adaptations for larger size and terrestrial habitus not seen in *Trichosurus*. The metacarpals of *Ngapakaldia* are much more rugose and relatively shorter and stouter (length of MC III < three times the width) than those of the arboreal possum. This is characteristic of all the vombatoids studied except *Thylacoleo* (see table 3). MC IV is slightly reduced in length, so that it is as short or shorter than MC II, whereas in *Trichosurus* it is clearly longer. The proximal ends of both MC III and MC IV are much higher (dorsoventrally) than wide, even though the carpal facets still have the same general shape as in *Trichosurus*. MC III has a longer (proximodistally) facet for MC IV, which creates a stronger area of contact between these two bones.

The greatest modifications are seen in metacarpals I and V. The first metacarpal is greatly reduced in length and broadened mediolaterally, so that the maximum width is greater than half the length (w/l = 0.6-0.7). While all vombatiforms show some reduction of this metacarpal, only in *Ngapakaldia*, wombats, and *Thylacoleo* is it reduced to this degree (*Thylacoleo* is unusual in that width exceeds length by 0.35). The proximal facet of MC I in *Ngapakaldia* is concavo-convex (medially convex, laterally concave), whereas in *Trichosurus* it is convex only. As in all the species studied, the shaft of the metacarpal is straight; it does not have the laterally concave curve seen in wombats.

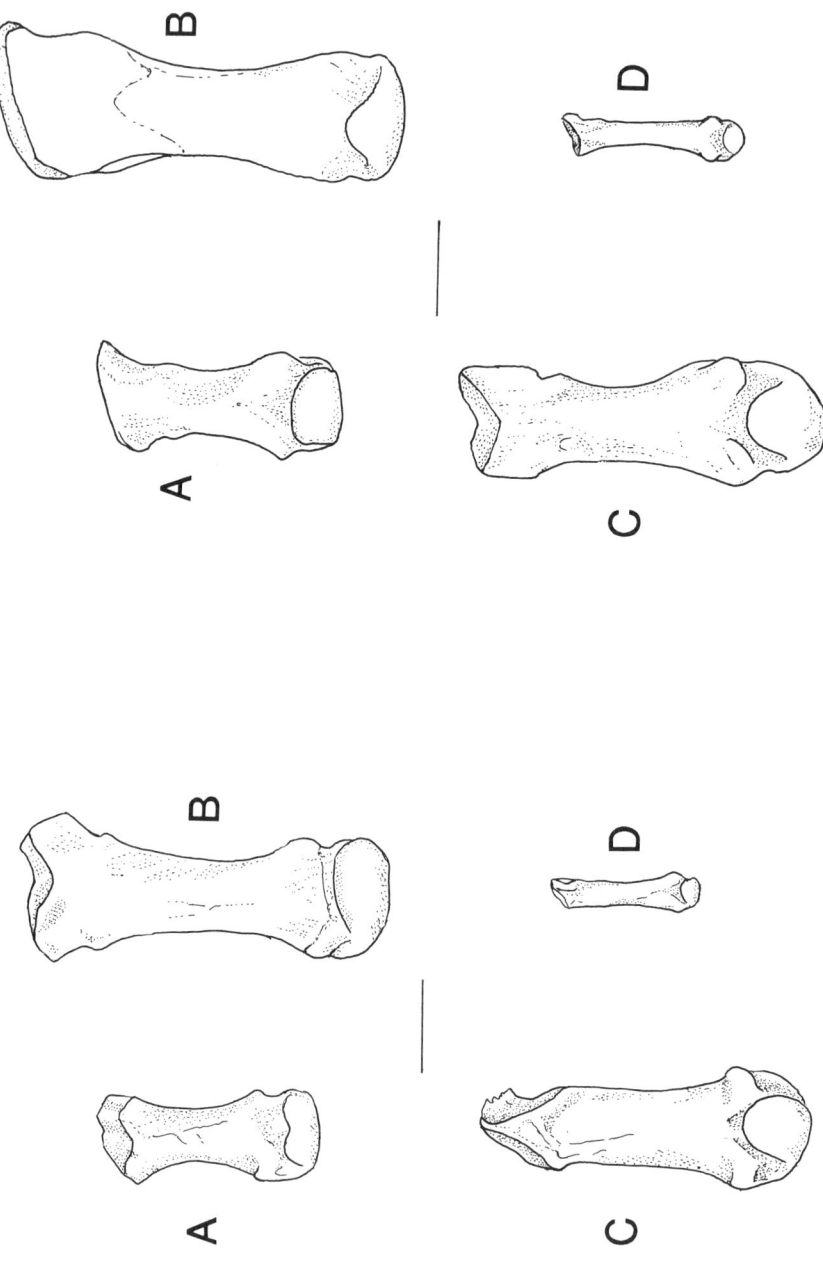

Fig 10. Metacarpal III, dorsal views, of A. *Ilaria*; B. *Lasiorhinus* (the hairy-nosed wombat); C. *Trichosurus*; and D. *Ngapakaldia*. Scale = 1 cm.

Fig. 9. Metacarpal II, dorsal views, of A. *Lasiorhinus* (the hairy-nosed wombat); B. *Ilaria*; C. *Ngapakaldia*; and D. *Trichosurus*. Scale = 1 cm.

Table 5. Apomorphies of the Ilariids and Vombatids

	Ilariids & Vombatids	Plesiomorphic state
1. Medial surface of scapula	Concave	Flat
*2. Thickness (ant-post) of ulna	> 0.2 length ulna	< 0.2 length ulna
*3. Distal facet of magnum	Fully concave	Convex or keeled
*4. Trapezoid facet of MC II	Faces posteriorly	Faces medially
*5. Proximal end of MC II	Expanded and concave	Mediolaterally compressed
Magnum facet of MC III	Convex	Grooved, concave
Outline of magnum facet	Semicircle	Triangle
MC II facet on MC III	Continuous with magnum facet	Separate, distinct, lateral
*6. Distal end of metapodials	Broad, flattened	Spherical
*7. Distal end of proximal phalanges	Dorsoventrally tapered, articulation for medial phalanx faces ventrally	Not tapered
*8. Unguals	Dorsoventrally flattened	Mediolaterally compressed
*9. Digit I of pes	Reduced, nonfunctional	Large, opposable
*10. Distal end of MT I	Convex, no condyles	Concave, with condyles (except *Diprotodon*)
*11. Proximal end of MT IV	Tapered	Square

(*) indicates weighted characters in the phylogenetic analysis

Metacarpal V is very similar to that of the wombats. It is very short and stout, the width at the distal end approximately equal to half the length of the shaft (in *Trichosurus*, w/l ≈ 0.34; in wombats, w/l ≈ 0.6). The proximal end is also broadened mediolaterally (width > dorsoventral height) giving a broader contact between MC V and the unciform, a characteristic of all the vombatoids studied (see table 3) except for *Lasiorhinus*. Also similar to *Vombatus* (but not *Lasiorhinus* or *Phascolonus*), the MC IV facet has a slight medial projection, and the shaft of the metacarpal is twisted, so that the distal condyle lies medial to the center of the unciform facet. But unlike wombats, the distal condyle is still spherical and unflattened, as in *Trichosurus*.

Phalanges

The proximal and medial phalanges of all the vombatiforms, including *Ngapakaldia*, are both shorter and stouter than those of *Trichosurus*, *Pseudocheirus*, and *Dromiciops*. The proximal phalanges of *Trichosurus* and *Dromiciops* are approximately as long as their corresponding metacarpals; in *Pseudocheirus*, they are longer. In most of the vombatiforms, the proximal phalanges are only 0.5 the length of their corresponding metacarpals, but in *Ngapakaldia* they are about 0.8 as long. The width of the proximal phalanges in the vombatiforms is greater than 0.2 of the length (*Ngapakaldia* ≈ 0.4), as opposed to less than 0.2 in the other animals.

The medial phalanges of the vombatiforms show even greater reduction (length < 0.5 length of corresponding MC) than the proximals, but again *Ngapakaldia* is the least reduced of the group (length med. phal./length MC ≈ 0.45). The length of the medial phalanges of *Dromiciops*, *Pseudocheirus*, and *Trichosurus* is greater than half the length of the corresponding metacarpal. The proximal and medial phalanges of *Ngapakaldia* are similar in transverse section and in the shape of their facets to *Trichosurus*. They are not dorsoventrally flattened as in wombats and *Ilaria*.

The unguals of *Ngapakaldia* are likewise mediolaterally compressed, as they are in all the species observed except wombats, *Ilaria*, and the spiny bandicoot *Echymipera kalubu*, in which they are dorsoventrally compressed. In *Diprotodon*, the unguals are more dorsoventrally flattened than in *Ngapakaldia*, but not so much as in wombats (table 5). The relative size of the unguals to the manus is comparable to that of *Trichosurus*; they are not disproportionately large as in the Pleistocene palorchestid *Palorchestes azael*.

In summary, the overall shape of the bones of *Ngapakaldia*'s manus maintains the primitive phalangeriform pattern, but with increased stoutness of the cuneiform, pisiform, metacarpals, and phalanges. The morphologies of the scapholunar, trapezium, trapezoid, and magnum are identical with those of *Trichosurus*. The shape of the facets of the metacarpals and phalanges, the rounded distal ends of the metacarpals, and the mediolaterally compressed unguals also appear to be plesio-morphic phalangeriform features. The broad pisiform, with its deep ulnar facet, the proximodistally deep cuneiform, and the wide fifth metacarpal (with the accom-panying broad MC V facet on the unciform) are synapomorphies of the terrestrial vombatoids (table 3). The short, broad metacarpal I and shorter phalanges are character states shared by all vombatiforms, including the arboreal koala (table 2). The increased stoutness and reduction in length of all the metacarpals is a shared trait of the diprotodontids, palorchestids, vombatids, and *Ilaria* (table 4).

PELVIC GIRDLE
(fig. 11)

There were two nearly complete pelves available for study. UCMP 69813 is the smaller of the two and articulates with the partial sacrum. The ilia are broken just anterior to the sacroiliac joint, but posterior to this point the pelvis is complete and in good condition. There is a nearly complete (dorsal corner broken) ?left epipubic bone. UCMP 72130 is missing only the anterior tip of the right ilium, but its overall shape is lopsided and appears to have been flattened compared to UCMP 69813. It has a ?left epipubic also, missing only the distal tip. There is also a broken left os coxa from a juvenile animal (UCMP 69084) which includes the ilium, acetabulum, and dorsal ramus of the ischium.

In all but a few features, the os coxae of *Ngapakaldia* are simply a larger version, again, of *Trichosurus*. The ilia of both are more compressed mediolaterally than dorsoventrally, affording a relatively smaller area for attachment of the iliacus, compared to that for the gluteus medius and erector spinae. The iliacus is a large strong muscle in the wombats, who use it to brace their hindlimbs against the digging action of the forelimbs (Elftman 1929), and accordingly their ilia are more dorsoventrally flattened.

Ngapakaldia shows a moderate increase in width of the iliac crest over that of *Trichosurus* (table 4). This flare is correlated with increased size of the erector spinae muscles, necessary to lift the anterior part of the body during the first phase of forward movement (ibid.). Like *Trichosurus*, the tip of the iliac crest does not extend as far laterally as in *Phascolarctos*, wombats, or *Diprotodon*. This lateral extension is a function of the width of the trunk (ibid.), indicating that *Ngapakaldia* was probably not as wide-bodied an animal as wombats or koalas.

Ngapakaldia's pelvis is also similar to that of *Trichosurus* in that, viewed posteriorly, the ventral rami of the ischii meet in an acute V-shape. Both species also have large obturator foramina (indicating less area for the surrounding muscles to attach) and a long pelvic symphysis. These features contrast with the short pelvic symphysis of wombats and the relatively small obturator foramina and U-shaped

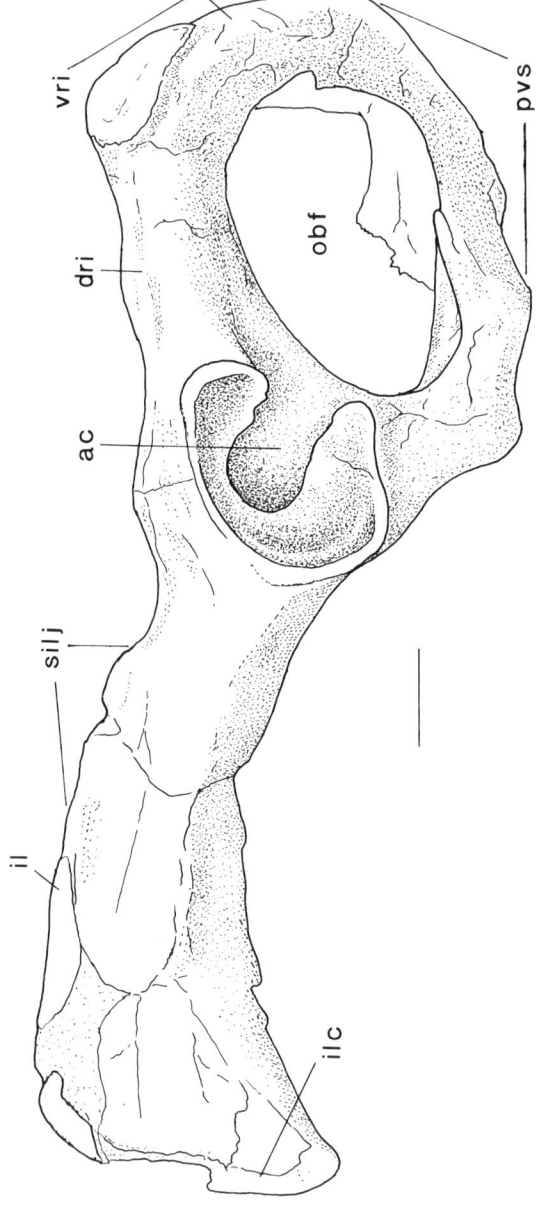

Fig. 11. *N. tedfordi*, pelvis, UCMP 72130, left lateral view. Scale = 2 cm. Abbreviations: ac, acetabulum; dri, dorsal ramus of ischium; il, ilium; ilc, iliac crest; obf, obturator foramen; pvs, pelvic symphysis; silj, sacroiliac joint; vri, ventral ramus of ischium.

Table 6. Apomorphies of the Palorchestids and Diprotodontids

	Palorchestids and Diprotondontids	Plesiomorphic state
1. Radial notch on ulna	Absent	Present
2. Depth (dorsoventral) of pelvis	≥ 0.8 med-lat width	< 0.8 width (except *Phas-colarctos*)
3. Epipubics	Thick	Thin

girdles of both wombats and koalas.

The location of the sacroiliac joint is slightly more posterior in *Ngapakaldia* than in *Trichosurus*. A similar though more extreme position in *Phascolarctos* and *Macropus* provides a larger area of origin for the erector spinae muscles (ibid.). The acetabulum of *Ngapakaldia* is slightly better-buttressed dorsally than in the arboreal species.

The length of the dorsal ramus of the ischium (compared to the length of the ilium) is most similar to that in *Thylacoleo* and only slightly shorter than in *Trichosurus*. The longer dorsal ramus of wombats gives a greater moment-arm to the hamstrings; the much shorter ramus of *Phascolarctos* brings the adductors closer to the femur, for more power of adduction rather than extension of the hip (ibid.). The in-between length of *Ngapakaldia* is consistent with an ambulatory pattern of locomotion, in which neither digging, running fast, nor climbing extensively play a dominant role.

The greatest difference between *Ngapakaldia* and *Trichosurus* is in the depth of the pelvis. The dorsoventral height at the symphysis in *Ngapakaldia* is almost equal to the width taken at mid-ischium (in *Trichosurus*, h/w ≈ 0.72). The only other species studied that come close to having this deep a pelvis are *Phascolarctos* and *Diprotodon*. In *Phascolarctos*, the deep pelvis is correlated with both the large volume of its viscera (especially the caecum and colon) and the more posteriorly positioned sacroiliac joint (which would decrease the size of the pelvic outlet without the increased depth). Since *Ngapakaldia* and *Diprotodon* have lophodont teeth (Archer 1984a), and therefore were browsing animals, it is likely that they too needed more depth to accommodate the lengthy abdominal viscera (table 6).

The complete epipubic bone (UCMP 69813) is identical in outline and length (relative to the os coxa) with that of *Trichosurus*, and very similar to those of the wombats. It is about two and a half times thicker (shortest dimension) than that in all the other species observed except *Diprotodon*, in which it is twice as thick as in *Ngapakaldia* (table 6). This implies a relationship to body weight, as the epipubics help support the abdomen.

HINDLIMB
(figs. 12 and 13)

There are several elements from the same adult animal (UCMP 72130): both femora, the left tibia, and both fibulae. They are nearly complete, except for the distal end of the left fibula, and in good condition, except for some erosion and deformation to the proximal end of the tibia. Also available for study is another complete left tibia in excellent condition (UCMP 69813) and a juvenile left femur (UCMP 60984) missing only the distal epiphysis.

All the bones of the hindlimb of *Ngapakaldia* are most similar in shape and proportions to those of wombats. Their femora (fig. 12) are equally broad proximally and thick in diameter at mid-shaft (compared to the length). The trochanter major is of the same shape and well developed in both animals. This contrasts with the slender, less developed femora of *Trichosurus* and *Phascolarctos*. The only differences between the femora in *Ngapakaldia* and wombats are that *Ngapakaldia* lacks the prominent lateral rugose area at the base of the trochanter major seen in wombats, and the patellar knob is much deeper (anteroposteriorly) on the medial side in wombats than it is in *Ngapakaldia*.

Like the femur, the fibula and tibia of *Ngapakaldia* (fig. 13) are stout and show adaptations for supporting greater weight on the ground. In *Ngapakaldia*, *Diprotodon* and wombats, the proximal ends of the fibulae are broader relative to length than in all other species studied. The facet for the flabella (fibular sesamoid) is expanded, and rises proximal to the tibial facet to give greater attachment area and mechanical advantage to the biceps femoris. Also, in these three species and *Thylacoleo*, the lateral notch on the distal end (for the digital flexors) is much shallower than in the arboreal species.

The tibia in all the vombatiforms studied (including *Ngapakaldia*) is about .20 shorter than the femur (in *Diprotodon*, it is .40 shorter). By contrast, the tibia is about the same length as (or longer than) the femur in the phalangeriforms, bandicoots, and dasyurids studied. The astragalar facet of the tibia in all the vombatoids has a deep pit posterior to the medial malleolus for the attachment of ligaments that greatly inhibit the amount of extension of the foot. *Trichosurus*, on the other hand, has a saddle-shaped astragalar facet, smoothly concave posterior to the malleolus, that allows the pes to invert around a branch. The anterior half of

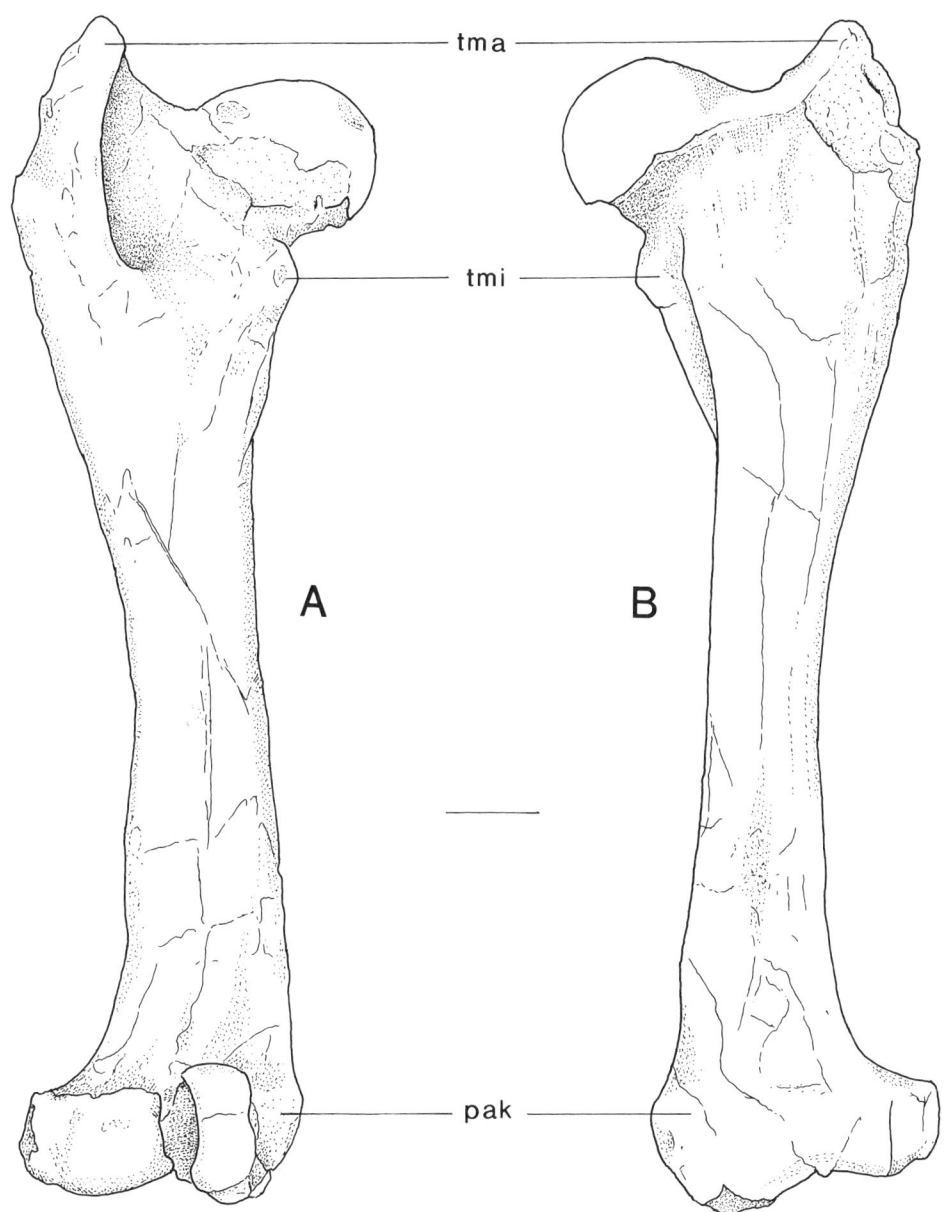

Fig. 12. *N. tedfordi*, left femur, UCMP 72130: A. posterior view; B. anterior view. Scale = 2 cm. Abbreviations: pak, patellar knob; tma, trochanter major; tmi, trochanter minor.

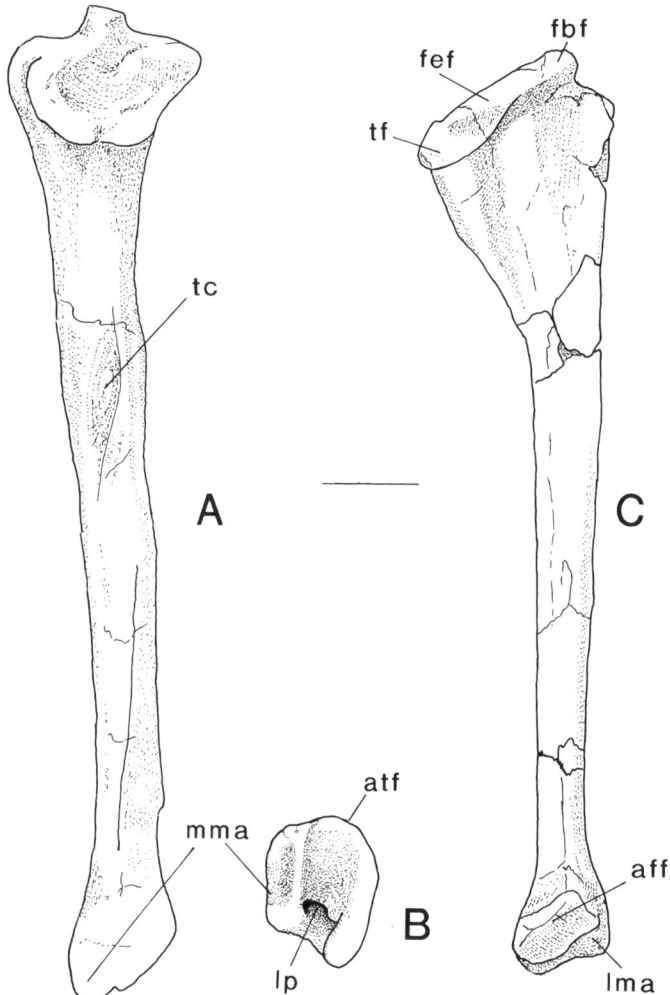

Fig. 13. *N. tedfordi*: left tibia, UCMP 69813: A. anterior view, B. distal view; C. right fibula, UCMP 72130, medial view. Scale = 2 cm. Abbreviations: aff, astragalofibular facet; atf, astragalotibial facet; fbf, flabellar facet; fef, femoral facet; lma, lateral malleolus; lp, ligamental pit; mma, medial malleolus; tc, tibial crest; tf, tibial facet.

this facet in *Phascolarctos* and *Ngapakaldia* is similar to that of *Trichosurus* (slightly saddle-shaped), but the posterior half is more convex and less sloping, especially in *Ngapakaldia*. The medial malleolus is also longer (anteroposteriorly) in *Ngapakaldia* than in either *Trichosurus* or *Phascolarctos*, but there is no deep indentation separating it from the medial astragalar facet, as there is in wombats.

In summary, the hindlimb shows a number of derived character states. The increased breadth of the proximal end of the femur, the reduced lateral notch on the distal fibula, and the less flexible tibioastragalar joint are features *Ngapakaldia* shares with the other terrestrial vombatoids (table 3). Even koalas, and thus all vombatiforms examined, have a relatively shortened tibia (table 2). The medio-laterally expanded proximal end of the fibula may be a synapomorphy between diprotodontids, palorchestids, and vombatids (table 4).

THE PES
(fig. 14)

As with the manus, all observations on the pes were made from casts of the holotype (UCMP 126722, left pes), which appears to have been in excellent condition. From the collection of disarticulated elements, only the second metatarsal, both phalanges of digit I, and the fourth ungual are missing. These bones are present in the cast of an articulated foot, but it is apparent that they have been reconstructed from fragments. Fortunately, the proximal end of MT II appears to have been genuine, as the shape of the mesocuneiform facet is a unique feature of *Ngapakaldia*. A new reconstruction of the articulated pes was made, as was the manus, by comparison to *Trichosurus*. The mesocuneiform, which was absent from the original recon-struction, was identified and included. Again, superimposed upon the general phalangeriform pattern are adaptations for terrestriality and increased size shared by the other vombatiforms.

Tarsals (fig. 15)

The astragalus and calcaneum "strongly reflect species-specific mechanics of the foot as well as the inherited genetic constraints often highly diagnostic of families, orders, or even supraordinal categories, in addition to being the most commonly found and easily identifiable postcranial elements of the mammalian fossil record" (Szalay 1984). There are distinct differences in the structure of these two bones between arboreal and terrestrial animals, and between different taxa.

The astragalus of *Ngapakaldia* (fig. 15), for example, is apparently that of a large terrestrial animal, with closest similarity to wombats and *Phascolarctos*, despite the general resemblance of the pes as a whole to the arboreal phalangers. In all the

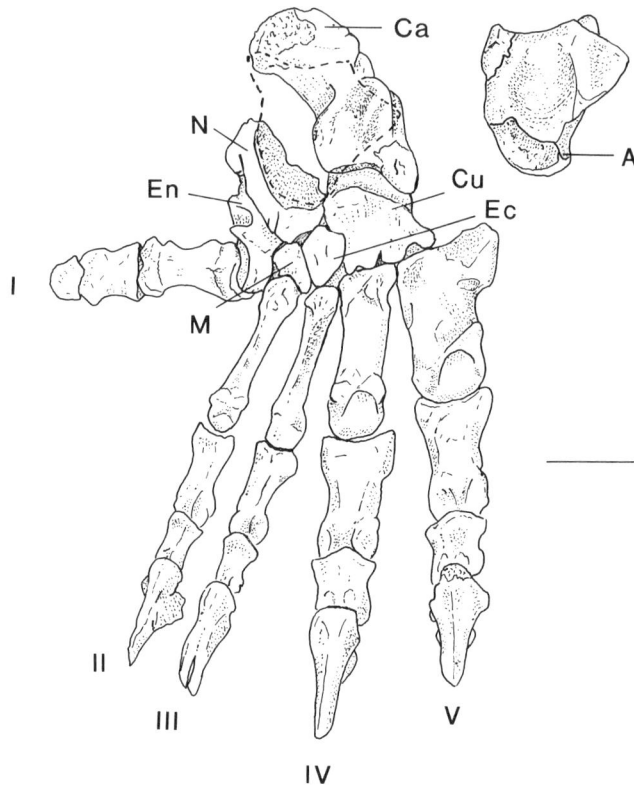

Fig. 14. *N. tedfordi*, left pes, UCMP 126722, dorsal view. The slender syndactylous digits and opposable first digit indicate an arboreal ancestry, but the trenchant astragalus is a synapomorphy of the terrestrial vombatiformes. Scale = 2 cm. Abbreviations: A, astragalus; Ca, Calcaneum; Cu, cuboid; Ec, ectocuneiform; En, entocuneiform; M, mesocuneiform; N, navicular.

vombatiforms studied, the lateral tibial facet is a concave trochlea, set between two anteroposteriorly trending ridges (table 2). The medial ridge, or tibial knob, separates the medial tibial facet from the concave lateral tibial facet, and the lateral ridge separates the lateral tibial facet from the fibular facet. This contrasts with the smoothly convex dorsal surface of the astragali in *Trichosurus*, *Phalanger*, and *Pseudocheirus*, which show no division between the tibial and fibular facets. The trochlea of *Ngapakaldia* is particularly deep, and comparable to that of *Vombatus*, creating a strong ankle joint that lacks the freedom of movement seen in phalangers. The medial tibial facet of *Ngapakaldia*'s astragalus is not a flat articular surface, as it is in wombats, but rather an indented ligamental pit as in *Phascolarctos*. The fibular

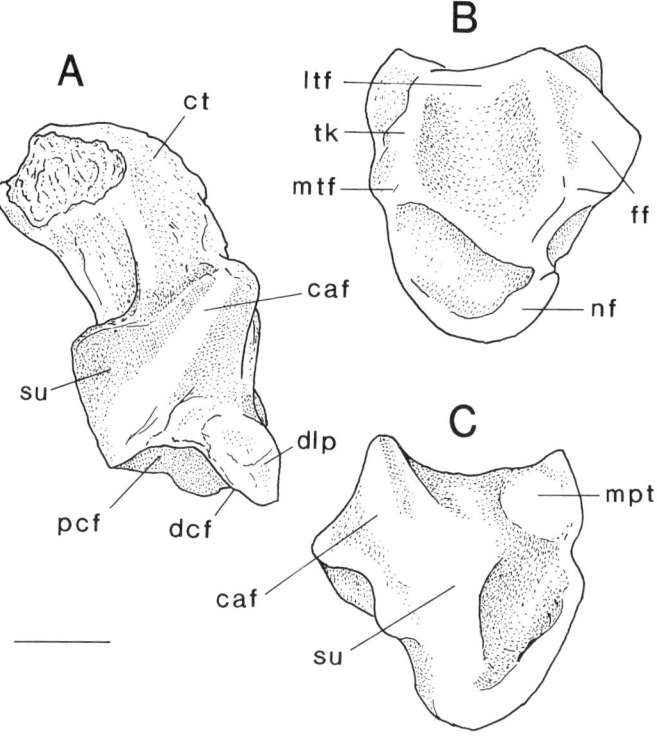

Fig. 15. The left astragalus and calcaneum of *N. tedfordi*, UMCP 126722: A. dorsal calcaneum; B. dorsal astragalus; C. ventral astragalus. Scale = 1 cm. Abbreviations: caf, calcaneoastragalar facet; ct, calcaneal tuber; dcf, distal cuboid facet; dlp, distolateral process; ff, fibular facet; ltf, lateral tibial facet; mtf, medial tibial facet; mpt, medial plantar tuberosity; nf, navicular facet; pcf, proximal cuboid facet; su, sustenaculum; tk, tibial knob.

facet is reduced compared to that of the phalangers, as it is in all the vombatiforms studied (table 2).

On the plantar side, the astragalus of *Ngapakaldia* shows more similarities to *Trichosurus*. The calcaneoastragalar facets of both are more deeply concave than in either the wombats or *Phascolarctos*, and the sustenacular facet nearly touches the medial plantar tuberosity. These features are separated by a deep sulcus in the wombats and *Phascolarctos*.

The general shape of the calcaneum is also more like that in *Trichosurus*, with a narrow projecting calcaneoastragalar facet and a long distolateral process that contacts the posterolateral side of the cuboid. In both of these animals and *Phascolarctos*, there is a prominent plantar tubercle on the ventral side of the proximal cuboid facet, and the length of the sustenaculum is more than .25 the length of the whole calcaneum. The major difference, however, between the arboreal species *Trichosurus* and *Phascolarctos* and the terrestrial *Ngapakaldia* is in the shape of the calcaneal tuber. As in the other large terrestrial animals, the tuber in *Ngapakaldia* is long and rugose. It curves strongly medially, so that the posterior end reaches even with an anteroposterior line through the medial side of the navicular.

The navicular bones of *Ngapakaldia* and *Trichosurus* are extremely similar: the astragalar facet is identically shaped in both, with a prominent angular bulge at the ventral edge of the cuboid facet. The rest of the bone is also nearly identical, although the posterodorsal corner in *Ngapakaldia* is rounded, as it is in all the other vombatiforms, not pointed as in *Trichosurus*.

In general outline, the cuboid of *Ngapakaldia* looks most like that of *Phascolarctos*, but slightly broader mediolaterally (compared to the proximodistal length) to accommodate the enlarged MT V facet. In all of the vombatoids studied, the MT V facet is as large as or larger than that for MT IV, whereas the MT IV facet is larger in the arboreal species like *Phascolarctos* and *Trichosurus*. The calcaneal facet of the cuboids in both *Ngapakaldia* and *Trichosurus* is separated into a flat proximal portion and a convex dorsolateral area that articulates with the distolateral process of the calcaneum.

The ectocuneiforms of *Trichosurus*, *Phascolarctos*, *Ngapakaldia*, and *Priscileo* are all fairly similar. They are nearly as long proximodistally as they are dorsoventrally high, with a strongly concave lateral side for the cuboid, a distolateral facet for MT IV, and a convex medial side for the mesocuneiform. The plantar tuberosity in *Ngapakaldia*'s ectocuneiform is large and rugose compared to the others, but it does not have the lateral protrusion seen in *Trichosurus*.

The mesocuneiforms of the two genera *Ngapakaldia* and *?Pitikantia* are unique. Of all the animals studied, only these early palorchestids have a saddle-shaped (vertically convex, mediolaterally concave) MT II facet. In all the others, it is concave only. The shape of the rest of the mesocuneiform in *N. tedfordi* is most similar to *Trichosurus*. It is slightly crescent-shaped when viewed dorsally: concave on the lateral side (ectocuneiform facet) and convex on the medial side (entocuneiform facet). It tapers posteriorly and appears to have had only minor contact with the navicular. A distolateral protrusion appears to have touched MT III; but the medial side is shorter, allowing MT II to contact the entocuneiform. The convex part of the MT II facet slopes ventrolaterally until it contacts the lateral side, forming one continuous anteroventral surface.

The entocuneiforms of *Ngapakaldia* and *Trichosurus* are identical in shape, being anteroposteriorly elongate, with a curly S-shaped appearance when viewed dorsally.

They both have a concavo-convex MT I facet, which indicates an opposable hallux, a long mesocuneiform facet (equal to or longer than the navicular facet), and well separated attachments for metatarsals I and II. *Ngapakaldia* has a small facet for MT II just anterior to the mesocuneiform facet; *Trichosurus* has only a point for ligament attachment.

Metatarsals

Metatarsals I, III, and IV are nearly identical in shape in both *Ngapakaldia* and *Trichosurus*, with *Ngapakaldia's* being larger and more rugose. MT I is a large weight-bearing bone with a concavo-convex proximal facet and two rounded condyles at the distal end. There is a greater concave component to the entocuneiform facet in *Ngapakaldia* (table 2) than in that of *Trichosurus* (which appears almost entirely convex). This feature would limit the motion of the joint in the vertical plane. When articulated, MT I sits at a right angle to MT II, just as it does in all the arboreal species studied.

The proximal end of MT III (ectocuneiform facet) is a convex triangle with a ventral apex. The proximal end of the lateral side has a flat facet and ligament scar for the articulation of MT IV.

MT IV in *N. tedfordi* is relatively wider for its length than that of *Trichosurus*, but has the same saddle-shaped proximal cuboid facet. This facet is completely convex in *N. bonythoni*. The MT V facet in both *Ngapakaldia* species is nearly flat, as it is in the other vombatoids studied, while it is concave in *Trichosurus*.

Metatarsal V is much shorter and broader in *Ngapakaldia* than in *Trichosurus*, with a very large lateral flange. Like the fifth metacarpal, it resembles its counterpart in wombats more than that of the arboreal possums (table 4). The cuboid facet is broader mediolaterally (table 3), and the MT IV facet is flat rather than convex as in *Trichosurus*. The increased breadth and stoutness of metatarsals IV and V indicate that *Ngapakaldia* must have put most of its weight on the lateral side of the foot. This is not unusual, considering that the animal still has an opposable hallux and typically syndactylous (unenlarged) digits II and III.

Since the early palorchestids *Ngapakaldia* and *Pitikantia* have a unique MT II facet on the mesocuneiform, they should also have the matching saddle-shaped mesocuneiform facet on MT II. This bone is unavailable for *Pitikantia*, but both *N. tedfordi* and *N. bonythoni* have mediolaterally compressed concavo-convex proximal facets on MT II. The facet also is oblique: it runs from the dorsolateral edge of the proximal surface to the medioventral corner. The shape of the facet allows a greater surface area for contact with the mesocuneiform, while increasing flexibility of movement both dorsally and medially.

Phalanges

What was said for the phalanges of the manus could be repeated for the phalanges of the pes. The length of the proximal and medial phalanges (compared to the corresponding metatarsal) in all the vombatiforms is considerably shorter than in the non-vombatiforms studied (*Trichosurus, Pseudocheirus, Dromiciops,* and three genera of dasyurids). In *Ngapakaldia,* the length of the proximal phalanges (prox. phal. length/MT length = 0.68) is less reduced than in the rest of the vombatiforms (ppl/MTl ≈ 0.4-0.6), but the relative length of the medial phalanges is equally short in *Ngapakaldia, Phascolarctos,* and the wombats (med. phal. length/MTl = 0.35-0.43). The proximal phalanges of the non-vombatiforms are nearly as long as the corresponding metatarsal, and the medial phalanges are at least half as long.

Proximal phalanges IV and V are quite broad in *Ngapakaldia, Diprotodon,* and the wombats (w/l > 0.40) compared to the non-vombatiforms studied (w/l ≤ 0.2), with the koala intermediate (w/l = 0.34). The relative width of the phalanges of digits I, II, and III in *Ngapakaldia* remains similar to those of *Trichosurus.*

The unguals, like those of the manus, are mediolaterally compressed, most similar to those in *Trichosurus, Phascolarctos,* and other non-fossorial species.

In summary, the pes shows a mixture of derived vombatiform and primitive phalangeriform traits. As in the manus, some bones (the navicular, ectocuneiform, and entocuneiform) are practically identical with those of *Trichosurus.* The mediolaterally compressed unguals and rounded distal condyles of the metatarsals are also plesiomorphic features. Most of the facets have the same shape as in the arboreal phalangeriforms, except for the more concave entocuneiform facet on MT I and the unique saddle-shaped facet between MT II and the mesocuneiform. The shape of the MT I facet is a synapomorphy for all the vombatiforms, as are the trenchant astragalus, with its decreased fibular facet, and the shortened phalanges (table 2). The increased MT V facet on the cuboid is a vombatoid trait, not shared with the arboreal koala (table 3). The increased stoutness of MT V and phalanges IV and V is seen in both the palorchestids and vombatids (table 4).

Ngapakaldia Bonythoni
(fig. 16)

Observations were made from casts and original material of the holotype (UCMP 57258). The casts are of the ulna; radius; carpus (excluding the trapezoid); metacarpals; proximal phalanges I, II, and IV; medial phalanx IV of the manus; and metatarsals I, II, III, and IV. The original tarsus minus the astragalus and calcaneum was also available.

Most of the elements are identical with those of *N. tedfordi* in shape, but average

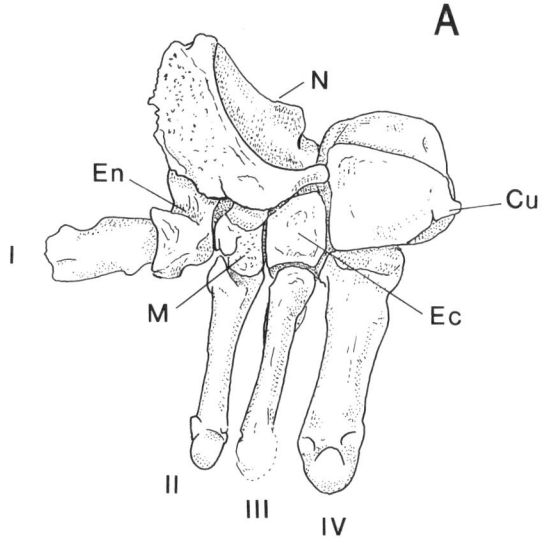

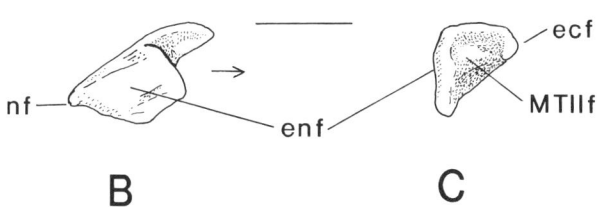

Fig. 16. A. *Ngapakaldia bonythonyi*, left pes, UCMP 57258, dorsal view; scale = 2 cm. B. Mesocuneiform, questionably referred to *Pitikantia dailyi*, UCMP 60981, medial view; C. mesocuneiform, *P. dailyi*, anterior view; scale = 1 cm. Abbreviations as in fig. 14, plus enf, entocuneiform facet; ecf, ectocuneiform facet; nf, navicular facet; MTIIf, metatarsal II facet.

0.3-0.4 larger. The length of the radius is only 0.17 longer, however, indicating that *N. bonythoni* had disproportionately short limbs compared to the size of its feet. The length of the olecranon process compared to the length of the ulna is also shorter in *N. bonythoni*, which is to be expected in a large animal. The extremely graviportal *Diprotodon* has no olecranon process at all to insure full extension of the joint on its columnar limbs.

Other features of *N. bonythoni* that differ from *N. tedfordi* include a larger hamate process on the unciform, a relatively longer metacarpal I, a more pronounced medial twist of the distal end of metacarpal V, a totally convex proximal end of metatarsal IV (*N. tedfordi*'s is concavo-convex), and a very differently shaped mesocuneiform.

The mesocuneiform of *N. bonythoni* has a flat posterior side that makes broad contact with the navicular, unlike that in *N. tedfordi*, which tapers to a point posteriorly. It also has a convex anterodorsal edge, rather than a concave one. The unique saddle-shaped MT II facet of *N. tedfordi* and *Pitikantia* (the most unique synapomorphy observed for the palorchestids) appears to also be present in *N. bonythoni*. The bone is somewhat eroded on the anterior and dorsal surfaces, but the proximal end of MT II is distinctly concavo-convex.

Pitikantia Dailyi

Elements of a left pes (UCMP 60981) questionably referred to this species (Stirton 1967) include a complete tarsus, metatarsals IV and V, and all phalanges of digits IV and V. The pes of *Pitikantia* is midway in size between *N. tedfordi* and *N. bonythoni*, the dimensions of the elements mentioned being 0.1-0.2 larger than those in *N. tedfordi* and about 0.2 smaller than those in *N. bonythoni*.

All the bones are identical in shape with those of *N. tedfordi*, indicating possible synapomorphies (terrestrial adaptations) and symplesiomorphies (similarities to *Trichosurus*) for the palorchestids, if indeed these postcrania are correctly associated with *Pitikantia*. The mesocuneiform has the distinctive saddle-shaped MT II facet with the concave anterodorsal edge, and tapers posteriorly to have minimal contact with the navicular. The only difference observed between this species and *N. tedfordi*, besides size, is that the astragalar surface of the cuboid is concave, as it is in wombats and *Diprotodon*, rather than convex. Such a minor feature as this, however, could very likely be an individual difference of this particular specimen.

DISCUSSION

It is apparent that *Ngapakaldia* is replete with plesiomorphic features inherited from an arboreal ancestor very similar in many aspects to the modern brush-tailed possum *Trichosurus vulpecula*. The structure of the vertebral column, pelvis, manus, and pes are particularly alike between these animals. Given the large, medio-laterally compressed unguals and the fully opposable hallux, it may even be possible that small members of the genus could have climbed trees on occasion (Rich, Van Tets, and Knight 1985), but, in my opinion, more like modern bears, due to the decreased flexibility of the ankle joint.

The modifications of the astragalus, calcaneum and distal tibia indicate that *Ngapakaldia* was primarily adapted to a terrestrial habitus. The shape of these bones and the proportions of all the limb bones show strong affinities to the terrestrial vombatoid group, and more particularly to the wombats. The trenchant dorsal surface of the astragalus, and the shortened proximal and medial phalanges of both manus and pes, are features held in common not only by the vombatoids, but also by the arboreal koala (table 2).

The broad median epicondyle and the long deltoid ridge of the humerus, the relatively short radius, and the long olecranon process of the ulna are features that Hildebrand (1988) associates with scratch-diggers. The morphology of the manus, however, would argue against such a conclusion, and leaves us wondering what *Ngapakaldia* did with its strong forelimbs without a fossorial manus like that of the vombatids. A late Pleistocene member of the palorchestid family, *Palorchestes azael*, shows powerful forelimbs with 12 cm. long, mediolaterally compressed unguals (Archer and Bartholomai 1978, Flannery 1983). It has been suggested (Flannery 1983; Rich, Van Tets, and Knight 1985) that this huge bull-size animal may have used its limbs to strip branches off trees or uproot bushes in search of tubers and roots. Perhaps *Ngapakaldia* showed a similar behavior, although on an appreciably smaller scale.

The structure of the mesocuneiform bone appears to be unique in *Ngapakaldia* and *Pitikantia*. The shape of the posterior end is flat rather than tapered in the larger *N. bonythoni*, but all three species show the unusual saddle-shaped MT II facet.

There are few shared derived features of the postcranial skeleton between these early palorchestids and the large graviportal diprotodontids, other than those related to size and terrestriality. Like the palorchestids, *Diprotodon* (the only diprotodontid with a complete pes) has an opposable hallux and greatly increased fourth and fifth digits of the pes without the accompanying increase in size of the syndactylous digits seen in the wombats. Both *Ngapakaldia* and *Diprotodon* must have borne the majority of their weight on the lateral side of the foot, although the palorchestids probably used the medial digits more, as evidenced by the increased flexibility of the mesocuneiform MT II facet.

ILARIA
Description and Comparisons

The single unique genus *Ilaria*, of the new family Ilariidae, consists of two species, *I. illumidens* and *I. lawsoni* (Tedford and Woodburne 1987), of which only the genotypic species *I. illumidens* has associated postcrania. The *I. illumidens* material was collected in 1973, during fieldwork conducted by R.H. Tedford and colleagues in the Namba Formation, Tarkarooloo Basin, South Australia. It is part of the Pinpa Fauna of medial Miocene age (ca. 14-17 Ma; Woodburne et al. 1985).

The similarity of these remains to vombatoids was immediately recognized, and they were designated "Vombatoidea genus B" (genus A being a smaller, as yet undescribed, wombat-like animal; Tedford et al. 1977). Though the cranium was somewhat crushed, enough of the basicranium was preserved to show that the squamosal made a major contribution to the auditory bulla as in other vombatoids. The Ilariidae differ from the other vombatoid families in the complex structure of their molar crowns, which show an incipient transverse lophodonty imposed upon a W-shaped ectoloph (dilambdodont pattern). They also possess a bulbous P_3 and a lower dentition unique among the Diprotodontia in retaining a tribosphenic trigonid on all lower molars. The retention of apparently primitive features, and the similarity of the upper dentition to certain phascolarctids, help support the argument that the dentition of the palorchestids and diprotodontids is secondarily simplified from the more complex tooth structure seen in *Ilaria*, as suggested by Tedford and Woodburne (1987).

The postcrania can be separated into two groups: those that are believed to belong to the same individual as the holotype cranium, and those that are referred to the species by their proximity in matrix. Associated with the holotype are an atlas, two thoracic vertebrae, one lumbar, the first sacral vertebra, and a left scapula. The referred specimens include a nearly complete right ulna and two fragments; a styloid process; a partial right radius; a right magnum; left metacarpals I, II, and III; three proximal phalanges, one medial and one ungual; a right metatarsal IV; a phalanx of digit I of the pes; and a fragment that might be a part of a syndactylous metatarsal.

Ilaria Illumidens, Holotype (AMNH 102649)

VERTEBRAL COLUMN

Atlas (fig. 17)

The atlas is the only cervical vertebra preserved. It is in good condition and nearly complete, with only a few millimeters of the tips of the transverse processes missing. It is a thick, massive bone, nearly identical in overall shape with that of the wombats, though twice the size.

As in the wombats, the anterior articular facets (fig. 17A) are broad and flaring, without the prominent dorsomedial projections seen in *Trichosurus* or *Phascolarctos*. In these two latter species, the intervertebral foramina are barely, if at all, visible when viewed anteriorly, whereas they are easily visible in *Ilaria*. A prominent alar notch (fig. 17C) is another feature that *Ilaria*'s atlas has in common with that of most wombats, although it tends to become less noticeable (covered by more bone) in older wombat specimens. Neither of these features is unique to these groups, as they are also evident in *Pseudocheirus* and *Dromiciops*.

Prominent medial projections (fig. 17B) extend from the posterior articular processes for attachment of the odontoid ligament that separates the neural canal from the odontoid cavity. This feature has an identical shape in wombats, is similar but rounder in *Trichosurus*, and is absent from the two koala specimens.

All vombatiforms studied, including *Ilaria*, have an incomplete ventral arch, but this feature is variable within the phalangeriforms, and Waters (1967) reports a *Vombatus* with a partially fused bony ventral arch. The ventral arch of *Ilaria* is more open than that of the wombats observed in this study, but shows a very rugose ventral surface, especially on the posterior half, presumably for the attachment of ligaments or cartilage that bridged the gap in life.

The dorsal arch of the atlas (fig. 17C) has a nearly uniform length (anteroposteriorly) all the way across, creating a very stout, sturdy structure. This feature is also seen in *Trichosurus*, *Phascolarctos*, *Diprotodon*, and one older specimen of *Lasiorhinus*.

The transverse processes are not constricted at their bases, a feature unique to *Ilaria*. Even though the tips of these processes are missing, it nevertheless appears that they were relatively short and not greatly expanded, a feature common to all diprotodontian species studied except *Diprotodon* and perhaps *Ngapakaldia tedfordi*. A complete atlas of *Ngapakaldia* was unavailable for comparison, but Waters (1967) reports that there is little to distinguish it from *Diprotodon* except size.

The atlas of *Ilaria* differs from that of the wombats in only two minor features. Wombats, especially older animals, have prominent bony projections extending

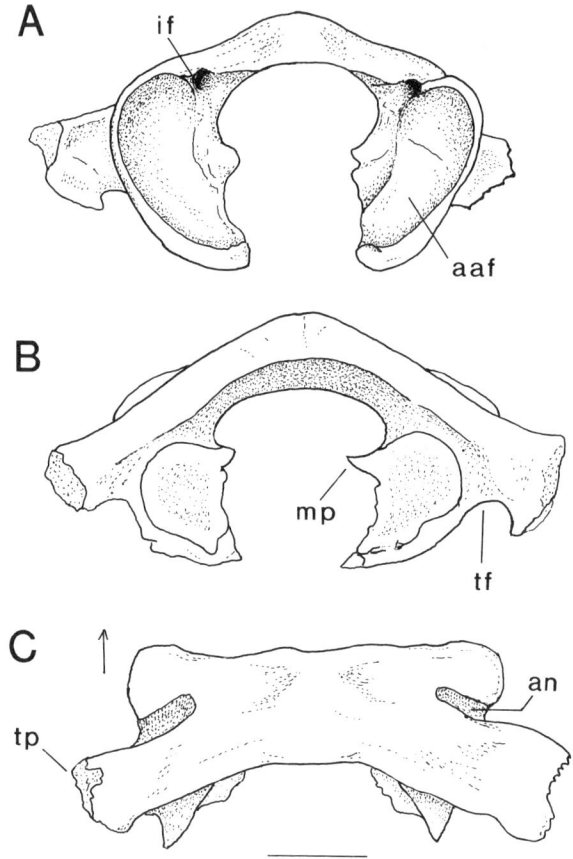

Fig. 17. *I. illumidens*, atlas, AMNH 102649: A. anterior view; B. posterior view; C. dorsal view. Scale = 2 cm. Abbreviations: aaf, anterior articular facet; an, alar notch; if, intervertebral foramen; mp, medial projection; tf, transverse foramen; tp, transverse process.

from the side of the ventral arch and the transverse process that nearly close off the transverse foramen. This is also seen in mature specimens of *Diprotodon* (Waters 1967). In *Ilaria*, *Trichosurus*, and *Phascolarctos*, these projections are represented only by tiny bumps or ridges. It is possible that this particular *I. illumidens* was a relatively young individual, for even though it has its full dentition, the molar crowns are only slightly worn (Tedford and Woodburne 1987). Wombats also have a relatively shorter atlas anteroposteriorly (length/height = 0.64-0.69), and the entire cervical series is greatly reduced in length in these animals. In *Ilaria*, the length-to-height ratio of this atlas (l/h = 0.76) is more similar to that of the three *Trichosurus* specimens examined (l/h = 0.74), perhaps indicating that although the atlas had the same broad strong connection to the skull as in the wombats, the animal probably had a longer neck.

In summary, the atlas is most similar to that of a young *Vombatus*, except for overall size and relative length. The similarity to wombats, however, may not be phylogenetically important, since none of the features cited is especially unique to either of these groups.

Thoracic vertebrae (fig. 18)

There are two thoracic vertebrae, estimated by comparison with the modern species to be approximately numbers four and eight, given a series of thirteen. While it is impossible to tell how many thoracic vertebrae there were originally, thirteen is the number found in all species studied with the exception of *Phascolarctos* (eleven) and *Vombatus* (fifteen). Both vertebrae are missing the transverse processes and most of their respective spinous processes. Facets for the head of the ribs, anterior and posterior articular processes, centra, and complete vertebral arches are present. The anterior and posterior surfaces of the centrum on the ?fourth vertebra are slightly eroded, and the left anterior articular process on the ?eighth is missing. The ?fourth vertebra has an anteroposteriorly shorter and dorsoventrally shallower centrum with a wider neural arch than the ?eighth, which has a deeper, more circular centrum (fig. 18B). Both vertebrae have well developed posterior demi-facets for the rib heads, but only the ?eighth shows anterior facets (fig. 18A). The anterior demi-facets of the ?fourth vertebra are either extremely faint or possibly eroded away.

The anterior articular processes of both vertebrae (fig. 18C) are small and flat, with the facets facing dorsally. This indicates their position well anterior to the anticlinal vertebrae which, in most animals, have more vertically oriented articular facets. The horizontal aspect of the facets would also facilitate lateral mobility of the vertebral column at these positions (Finch and Freedman 1986).

The overall shape of the vertebrae is similar to that of the other species studied, except in the dimensions of the centra. In *Ilaria*, like wombats, the greatest dimension of the centra is width; the least, height (width > length > height). Yet the centra of *Ilaria* are not quite as flattened as those of wombats, where height/width ratios range from approximately 0.36-0.66 among the thoracic verte-brae. The floor of the neural canal is flat in *Ilaria*, rather than concave as it is in wombats. The ratio of height to width of the centrum in the ?eighth vertebra is approximately 0.80, making the centrum appear almost circular in cross-section. This is only slightly deeper than in *Trichosurus* (0.6-0.7), and is comparable to that of

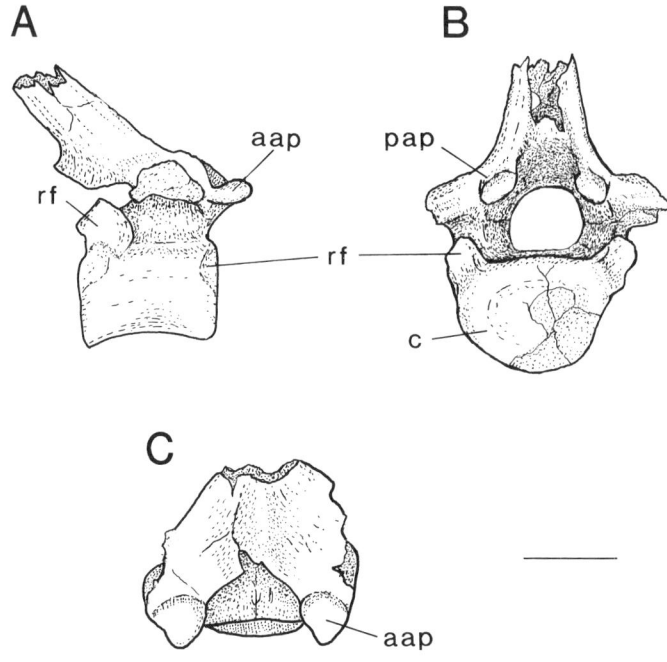

Fig. 18. *I. illumidens*, thoracic vertebrae, AMNH 102649: A. ?eighth thoracic vertebra, left lateral view; B. ?eighth thoracic vertebra, posterior view; C. ?fourth thoracic vertebra, dorsal view. Scale = 2 cm. Abbreviations: aap, anterior articular process; c, centrum; rf, rib facet; pap, posterior articular process.

the eighth vertebra of *Thylacoleo* and a vertebra (?sixth) of *Ngapakaldia*. However, the anteroposterior length of the centrum is the smallest dimension in *Thylacoleo* (short vertebrae, w > h > l), and the largest in *Ngapakalkia* (long, spool-shaped vertebrae, l > w > h), with *Ilaria* being intermediate.

Lumbar vertebra (fig. 19)

The single lumbar vertebra of *Ilaria* appears to be from an anterior position in the series (numbers 1 or 2) because of the small size of the centrum compared to that of the first sacral vertebra, assuming a plesiomorphic number of six lumbar vertebrae are present. The right transverse process and the posterior articular processes are completely missing, and there remain only about two centimeters of the bases of the left transverse process and the neural spine. The anterior articular facets are

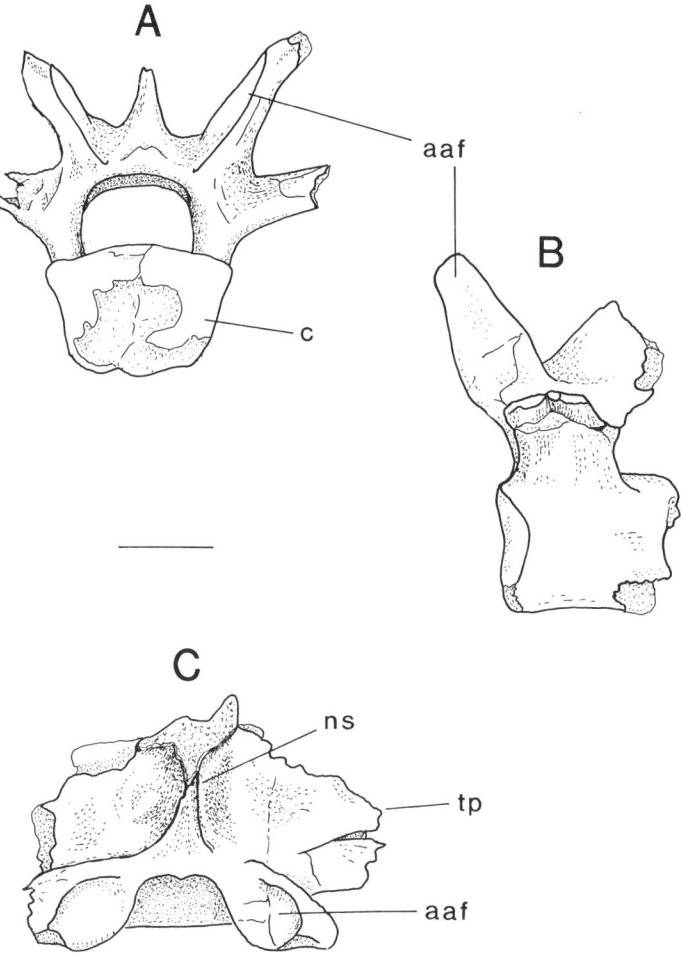

Fig. 19. *I. illumidens*, lumbar vertebra, AMNH 102649: A. anterior view; B. left lateral view; C. dorsal view. Scale = 2 cm. Abbreviations: aaf, anterior articular facets; c, centrum; ns, neural spine; tp, transverse process.

present, with the right mammillary process broken above the facet. The posterior surface of the centrum is partially eroded, but the rest of this structure and the vertebral arch are complete. There is only a small incipient accessory process posterior to the left transverse process (this area is missing on the right), which tends to support the idea that this is one of the first lumbar vertebrae. On specimens of extant diprotodontian taxa the accessory processes are smallest on the first and last lumbar vertebrae, but the size of the centrum here precludes the latter possibility.

The overall shape of this vertebra is most similar to that in *Thylacoleo*. The centrum is deep (fig. 19A), the height-to-width ratio (approx. 0.75) comparable to that in *Thylacoleo* and *Trichosurus*. However, the length of the centrum in *Thylacoleo* (compared to the height and width) is shorter than in *Ilaria*, and the centrum is considerably longer in *Trichosurus*. In both *Ilaria* and wombats, the length of the centrum is the greatest dimension with the height least; but in wombats, the height is so reduced that the centra appear broad and flattened in transverse section. In wombats, the neural processes of all but the most posterior lumbar vertebrae tend to extend straight up or tilt anteriorly. In *Ilaria*, the base of the neural process indicates that it tilted posteriorly, as they do in *Thylacoleo*. The transverse processes extend horizontally, as they do in all the vombatiforms studied except *Lasiorhinus*, but the thickness at their base indicates that they were much stouter (and perhaps broader transversely) than any species studied, excluding *Diprotodon*.

The most distinctive feature of *Ilaria* is the position of the anterior articular processes and facets. Viewed laterally, the processes extend upward at an angle of 60° to the anteroposterior axis of the centrum (fig. 19B). On all other specimens studied this angle is never more than 30° except again in *Thylacoleo*, where it is between 50° and 60°. This feature, combined with the relatively shortened length of the centra, seems to indicate that *Ilaria* was stocky in the lumbar region, with a spine that was relatively incapable of a great deal of flexion.

Sacral vertebra no. 1 (fig. 20)

Most of the first sacral vertebra is present, including the anterior articular facets, the anterior portion of the centrum, and parts of the wings (alae). The rest of the sacrum is missing, and the posterior and lateral portions of this vertebra are badly eroded. The neural spine is broken off, and the auricular surfaces for the ilium are missing on both sides, although the ventral portion of the wing is present. The posterior portion of the vertebra is missing just posterior to the base of the neural spine, resulting in no sacral foramina being present.

The centrum is nearly circular in transverse section but flat dorsally, on the floor of the neural arch. Its width is only slightly greater than the dorsoventral height, the same as in the lumbar and ?eighth thoracic vertebrae. The anterior articular facets are large, flat, and circular, making an open 45° angle with the horizontal plane. The anterior edges of the sacral wings bend forward sharply from the pedicles, the tip of an unbroken portion on the right reaching past the edge of the centrum (fig. 20B).

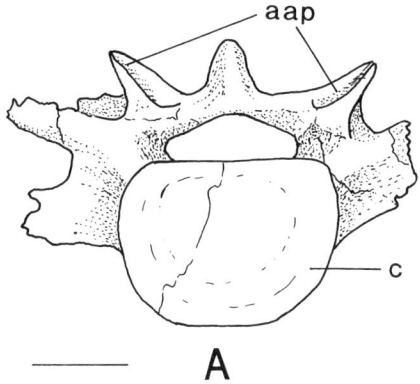

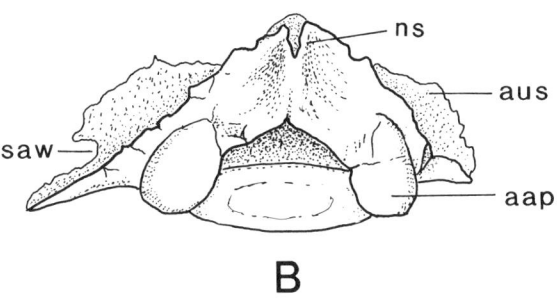

Fig. 20. *I. illumidens*, first sacral vertebra, AMNH 102649: A. anterior view; B. dorsal view. Scale = 2 cm. Abbreviations: aap, anterior articular processes; aus, auricular surface; c, centrum; ns, neural spine; saw, sacral wing.

It is difficult to make comparisons of the entire sacrum, since so much of it is missing. The depth of the centrum and the sharp anterior curve of the wings lateral to the articular processes make it appear that the sacrum was quite narrow and deep, as in *Thylacoleo*. The overall shape of the vertebra again most closely resembles that of the first sacral vertebra in the Pleistocene thylacoleonid. The sacra of the other vombatiforms, especially wombats, have broad flaring wings, relatively shallow centra, and more vertically oriented anterior articular facets. This would seem to indicate that, of all the species studied, only *Ilaria* and *Thylacoleo* had narrow hips, with little dorsoventral mobility between the sacrum and the last lumbar vertebra.

SCAPULA (figs. 21 and 22)

The only other postcranial bone associated with the holotype cranium is a nearly complete left scapula. The glenoid socket, the anterior border and supraspinous fossa, most of the spine, and most of the infraspinous fossa are intact. The coracoid process is slightly eroded, as is the area posterior to the glenoid cavity for attachment of the triceps muscle. The acromion process is broken even with the edge of the glenoid, so that it is impossible to determine how much of it extended ventral to the scapula. The dorsal and posterior borders of the blade are broken, but from the curvature of the intact portions it does not appear as if much of the scapula is missing from these edges.

The shape of the scapula is different from that in any of the other species studied. Unfortunately, however, there was no scapula available from *Ngapakaldia* for comparison. In general, it is most similar to those in *Trichosurus* or *Phascolarctos*, but with distinctive differences from either one. Like *Trichosurus*, the spine bisects the blade into two nearly equal halves, and the anterior border bulges forward at the middle of the blade (fig. 21). The bulge is greater in *Ilaria*, however, creating a deep concave curve on the anterior border of the neck. The posterior border is more like that seen in *Phascolarctos*, in that there is no concave curve at the neck; rather it bends convexly, affording a greater area for attachment of the teres minor muscle, and perhaps the infraspinous and triceps as well. This bend occurs directly oppposite the concave curve on the anterior border, making the entire blade tilt anteriorly relative to the glenoid cavity and neck. The spine follows this bend, curving gently forward. In all other species examined, the scapula tilts more posteriorly and the spine extends posterodorsally in a straight line.

In common with all species studied except wombats, the spine is perpendicular to the blade and rises to no more than 0.25 the length of the scapula, where it reaches its greatest height at the neck (fig. 22A). The lateral edge of the spine shows considerable overhang posteriorly, even more than in the wombats. This overhang is broken ventrally, so that it cannot be determined how wide the acromion process might have been.

As in wombats, the subscapular surface in *Ilaria* is strongly concave medially along the long axis (fig. 22A). The coracoid process is short and does not extend medial to the glenoid fossa as it does in *Trichosurus* or *Phascolarctos* (fig. 22A). This reduction may be an adaptation to terrestriality (the coracobrachialis and biceps brachii muscles, used in climbing, attach here), and is also seen in all the terrestrial vombatoids. Another feature of large terrestrial animals, shared by the scapulae of *Ilaria*, wombats, *Diproton*, and thylacoleonids, is a large rugose area on the posterior border just dorsal to the glenoid cavity for the attachment of the triceps (fig. 22B). This area is especially large in *Ilaria*, equal to approximately 0.25 the length of the entire scapula.

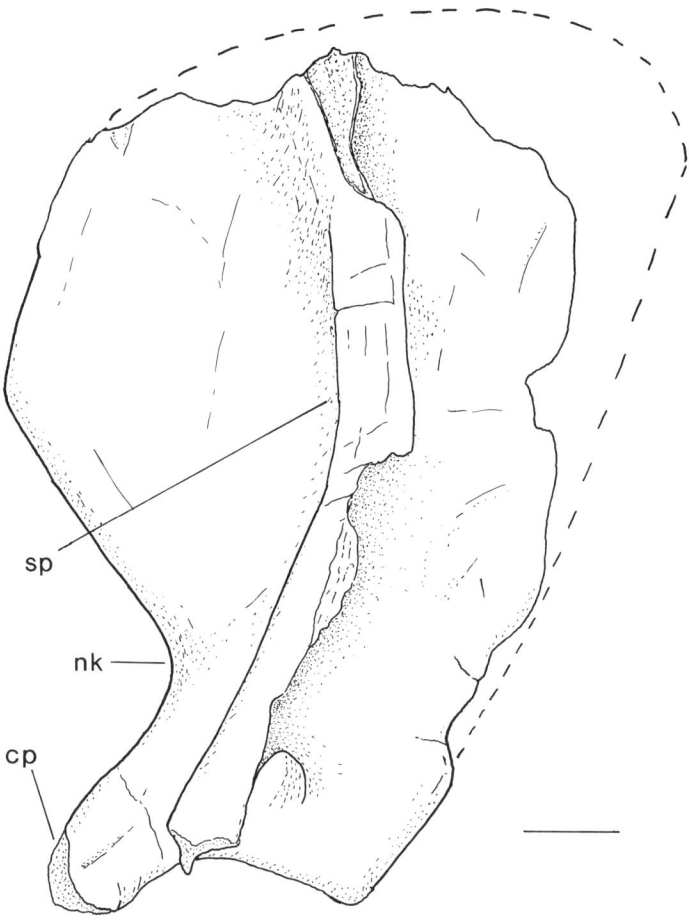

Fig. 21. *I. illumidens*, left scapula, AMNH 102649, lateral view. Scale = 2 cm. Abbreviations: cp, coracoid process; nk, neck; sp, spine.

The large supraspinous and infraspinous fossae, the strong, overhanging scapular spine, and the large triceps muscle scar all indicate an animal with powerful forelimbs. It is unfortunate that the teres major process is missing; but judging by its prominence in all species studied except *Diprotodon*, it is probable that it was quite angular, as shown by the dotted lines in fig. 21.

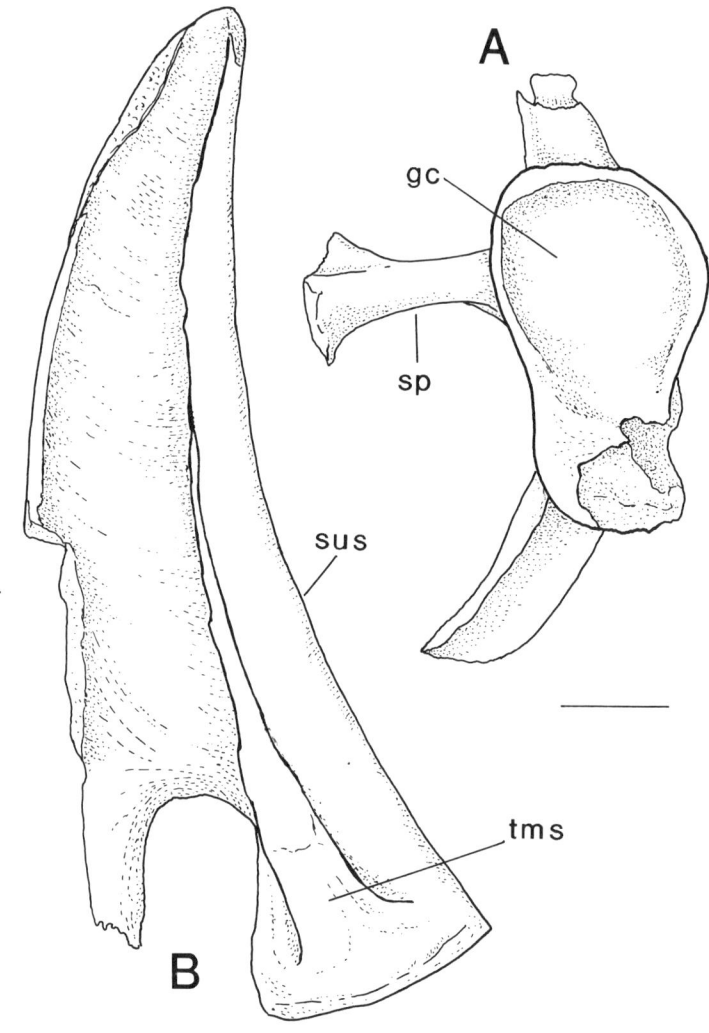

Fig. 22. *I. illumidens*, left scapula, AMNH 102649: A. distal view; B. posterior view. Scale = 2 cm.
Abbreviations: gc, glenoid cavity; sp, spine; sus, subscapular surface; tms, triceps muscle scar.

Ilaria Illumidens, Referred Material

FOREARM

Ulna (fig. 23)

There is one nearly complete right ulna (AMNH 102636), two fragments (AMNH 102651 and 102221), and an isolated styloid process (AMNH 102644). AMNH 102636 is broken just above the distal condyles, and is therefore missing the distal end including the styloid process. The proximal end is eroded so that it is impossible to determine the size of the anconaeus process and whether it might be enlarged as in wombats or lacking altogether as in *Diprotodon*. The area lateral to the semilunar notch is also eroded, removing the radial facet. However, half of this facet is present on AMNH 102221, which also includes part of the semilunar notch and about 11 cm. of shaft distal to this feature. AMNH 102651 is another fragment of a right ulnar shaft, about the same size as AMNH 102221, broken just below the semilunar notch.

The shaft of AMNH 102636 curves medially and is somewhat flattened medio-laterally. The greatest thickness (anteroposteriorly) is at the coronoid process, which extends to more than twice the thickness of the more slender shaft and olecranon process. In overall shape the ulna appears most similar to that of the wombats, except for the length of the olecranon process. This process equals 0.25 the total ulnar length in wombats, approximately 0.20 in *Ilaria* and *Ngapakaldia bonythoni*, and only about 0.10 in the other diprotodontians studied (with the exception of *Diprotodon*, which has essentially no olecranon). Interestingly, in the bandicoots, which are scratch-diggers, this feature varies in proportion between species, from 0.25 to 0.20 the total length. Hildebrand (1988) lists a number of expert scratch-diggers with an olecranon/ulnar length ratio of only 0.17 or more.

Similar to wombats, *Ilaria* has a very large coronoid process below the semilunar notch (fig. 23A). Presumably this is to help prevent disarticulation of the elbow joint in the flexed position, implying that, like wombats, *Ilaria* may be applying a fair amount of pressure to this joint when flexed. Because the anconaeus process is broken off, it cannot be determined whether or not this structure was also enlarged as in wombats. If *Ilaria* was only scratch-digging to a limited extent (like bandicoots), and not burrowing like wombats, then a large anconaeus, the purpose of which may be to prevent disarticulation of the elbow in the extended position, would not be needed.

The size of the ulnar elements AMNH 102636 and 102221 seems disproportionately small when *Ilaria* is compared to *Ngapakaldia bonythoni*, the largest of the palorchestids studied. The length of the molar tooth row in the

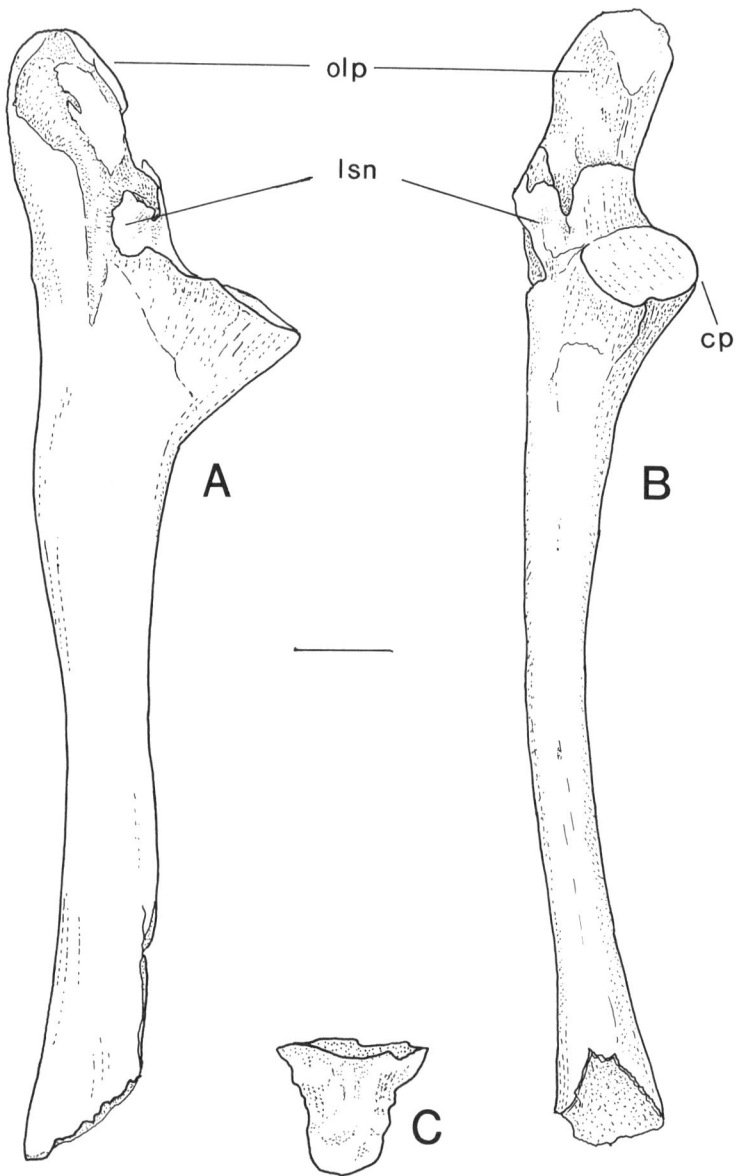

Fig. 23. *I. illumidens*: A. and B. right ulna, AMNH 102636; C. styloid process of ulna, AMNH 102644. Scale = 2 cm. Abbreviations: cp, coronoid process; lsn, lateral facet of semilunar notch; olp, olecranon process.

holotype *I. illumidens* (AMNH 102649) is greater (81 mm.) than that of *N. bonythoni* (73 mm.), but these *Ilaria* ulnae would have been slightly smaller (shorter and more slender) than the ulna of *N. bonythoni* (UCMP 57258), the same specimen from which the tooth row measurements were taken. The isolated styloid process, AMNH 102644, however, is slightly larger than the distal end of the ulna in *N. bonythoni*. This could imply that the ulnae referred to *I. illumidens* belong to smaller individuals, females (if sexually dimorphic), or juveniles, or possibly to some other species. A fourth, less likely, possibility is that *Ilaria* had a big head and small legs, and the styloid process belongs to a different, larger species.

Radius (fig. 24)

AMNH 102185 is a nearly complete shaft of a right radius with both ends missing. The proximal end with the radial head is broken just above the radial tuberosity; the distal end is broken just above the styloid process. Both the proximal and distal ends are oval in transverse section and anteroposteriorly compressed. The middle of the shaft is triangular in cross-section, since the posterior side is flat from the radial tuberosity to the broken distal end. The lateral edge of the shaft forms an angular ridge and has muscle scars where it would come in contact with the ulna; the medial side is more smoothly rounded, except for a small muscle scar halfway down the side. The shaft has a slight medial flexion and an eroded indentation on the posterior side of the proximal end that marks the position of the radial tuberosity.

The radius is most similar in both shape and size to that of *N. bonythoni*, which is consistent with the interpretation that the ulna (AMNH 102636) is from a smaller animal than *N. bonythoni* (the radius and ulna of *N. bonythoni*, UCMP 57258, are both from the same individual). As in all vombatoids, the radius has relatively broad and thick distal ends when compared to arboreal taxa such as *Phascolarctos* or phalangers, but it lacks the strong curvature of the shaft seen in wombats.

THE MANUS

There are a number of elements of the manus, AMNH 102636, from both the right and left sides, from the same individual as the ulna described above. Metacarpals II and III are about .50 larger than the corresponding elements in *Vombatus*, but the size of the molar tooth row of the holotype *I. illumidens* is twice the size of *Vombatus*. That there is no evidence of incomplete growth at the epiphyseal plates on the metacarpals makes it most likely that this was a relatively smaller adult animal (compared to the holotype), or that *Ilaria* had a larger head relative to the feet than *N. bonythoni*. The manus elements include a right magnum and meta-

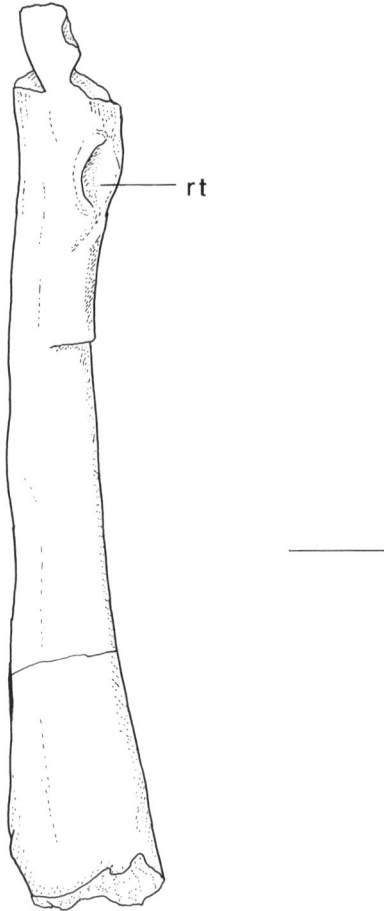

Fig. 24. *I. illumidens*, right radius, AMNH 102184, posterior view. Scale = 2 cm. Abbreviation: rt, radial tuberosity.

carpal II; left metacarpals I, II, and III; three proximal phalanges, one middle phalanx, and an ungual. All bones are in good condition and complete, except for some erosion of the proximodorsal surface of MC III.

Magnum (fig. 25)

The magnum of *Ilaria* is unique among the specimens observed. It is a thick wedge-shaped chunk of bone with a square, concave distal facet for metacarpal III, and a flat unciform facet. In comparisons of height to width, *Ilaria*'s magnum is most

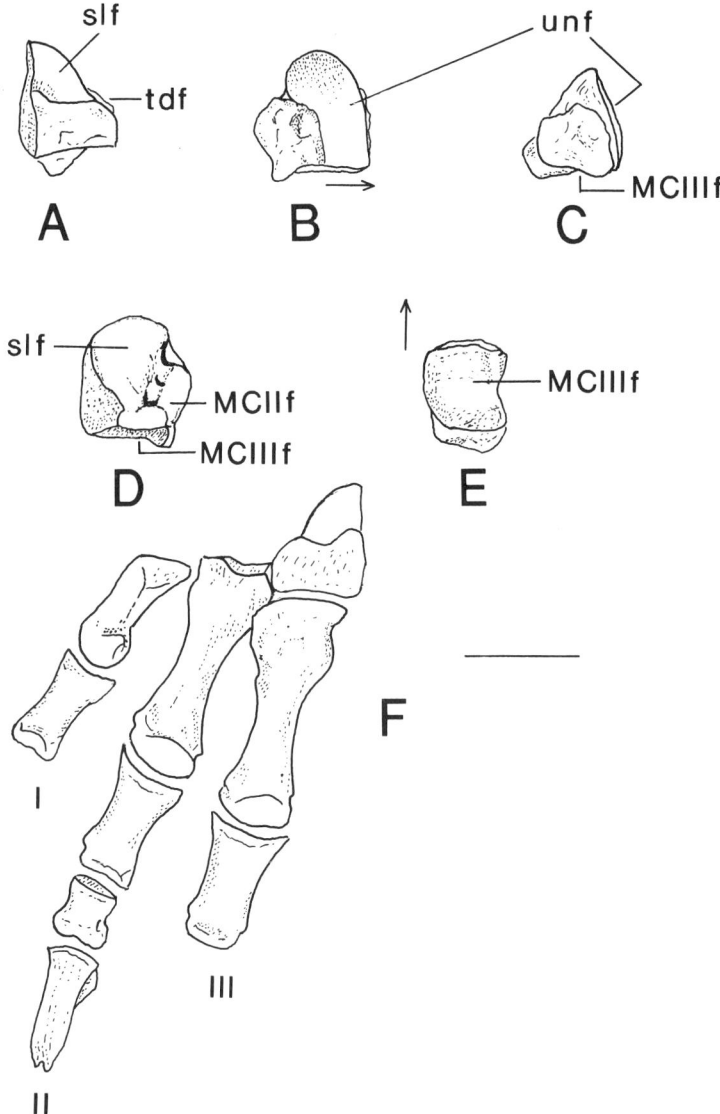

Fig. 25. *I. illumidens*, right magnum and elements of left manus, AMNH 102636: A. anterior view; B. lateral view; C. posterior view; D. medial view; E. distal view of magnum; F. left manus, dorsal view (with mirror image of right magnum added). Scale = 2 cm. Abbreviations: f, facet; MC, metacarpal; slf, scapholunar facet; tdf, trapezoid facet; unf, unciform facet.

similar to those of *Ngapakaldia* and *Trichosurus* (slightly higher proximodistally than wide), but the proximal projection that articulates between the scapholunar and unciform is much thicker in *Ilaria*, rather than compressed mediolaterally as in nearly all the other species observed (fig. 25A). *Lasiorhinus* approaches *Ilaria* in the shape of this projection but, like the other wombats, its magnum is wider than it is high.

While only in *Ilaria* does it have a square distal surface (it is roughly triangular in all others), the smooth, fully concave MC III facet (fig. 25E) is shared by wombats and the Miocene thylacoleonid *Priscileo pitikantus*. This feature gains greater significance when the extreme similarity between the metacarpals of *Ilaria* and those of the wombats are considered (see below). The flat unciform facet of *Ilaria* is unusual, because in all the other species examined there is an interlocking concavo-convex articulation between these bones which helps hold them together (fig. 25B). There is a deep pit ventral to the facet which would have held the ligaments that connected these bones, but the articulation still seems weak when compared to the other species.

Metacarpals (fig. 26)

Ilaria's left first metacarpal (fig. 26A) is most similar in shape and dimensions to those of *Trichosurus* and *Phascolarctos* (plesiomorphic), except for the shape of the distal condyle, which is more wombat-like. Its length is approximately half that of the third metacarpal, as in *Trichosurus*, whereas it is relatively shorter in the other vombatiforms. As in both *Trichosurus* and *Phascolarctos*, the shaft has a concave ventral curve, and the proximal facet for the trapezium is triangular and convex. The distal end, however, is slightly flattened dorsoventrally and broad, similar to the distal condyles of MC II-V of wombats and the koala.

Metacarpals II and III of *Ilaria* (fig. 26B, C) are distinctive in their remarkable same flattened, broad distal ends, and identically shaped proximal facets. Only in *Ilaria* and wombats is the proximal end of the second metacarpal squared-off, with a large posteriorly facing trapezoid facet and a sharply delineated, large, rectangular magnum facet, rather than tapered mediolaterally. The magnum facet is clearly separated from the concave MC III facet distal to it, and projects laterally over the medial corner of MC III as it does in wombats, and to a lesser degree in *Trichosurus*.

The magnum facet of MC III is also uniquely convex only in *Ilaria* and wombats, being slightly concave in all other species studied. (There is no MC III available from *Priscileo*, but since it has a concave distal facet of the magnum, it too may have a similar proximal surface of MC III.) The shape of the proximal end of MC III in *Ilaria* is a convex semicircle, identical with that in *Vombatus*, but the unciform facet

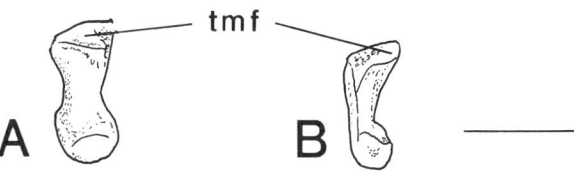

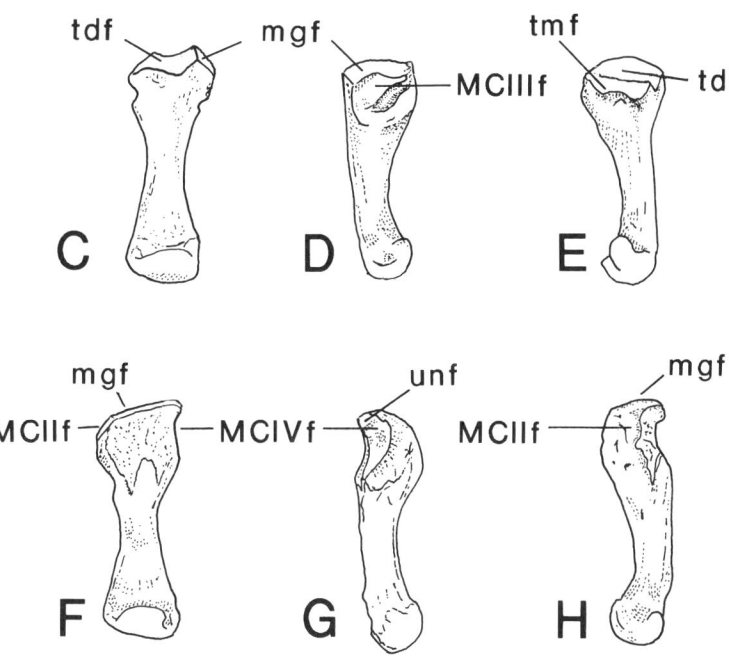

Fig. 26. *I. illumidens*, left metacarpals, AMNH 102636: A. dorsal view and B. lateral view of MC I; C. dorsal view, D. lateral view, and E. medial view of MC II; F. dorsal view, G. lateral view, and H. medial view of MC III. Scale = 2 cm. Abbreviations: f, facet; MC, metacarpal; mgf, magnum facet; tdf, trapezoid facet; tmf, trapezium facet; unf, unciform facet.

of *Ilaria* lacks the large lateral extension that projects proximally to MC IV in that taxon. Rather, the unciform articulation is reduced, like that of *Lasiorhinus*, whose manus in many ways does not appear quite as strong or interlocking as that of *Vombatus*. The surface of *Ilaria*'s metacarpals is very rugose, as are those of the wombats and the large diprotodontids and palorchestids (table 4).

Phalanges (fig. 27)

There are three proximal phalanges of the manus: the largest most likely from the third digit, a slightly smaller one from digit II or IV, and the smallest from the left digit I (fig. 27A). The proximal phalanges are tapered, flattened dorsoventrally at the distal end, the same as in wombats. In relative length, however (compared to their corresponding metacarpals), they are longer than in wombats, but shorter than either species of *Ngapakaldia*.

The medial phalanx (fig. 27B) is also dorsoventrally flattened like that of a wombat, so that the proximal facet faces posterodorsally instead of just posteriorly. The relative length may be even less than that of wombats (there is some uncertainty about which digit this bone represents), giving *Ilaria* very short, wide toes. The ungual claw is identical with those of wombats, being compressed dorsoventrally rather than mediolaterally, as in all the other species observed (fig. 27C). Only in wombats and *Ilaria* is the proximal facet of the ungual consistently wider than it is high (dorsoventrally). Also, the distal ends in both species contain vascular perforations. Owen (1874) interpreted these perforations in modern wombats as necessary for "quickly worn digging claws."

In summary, the forelimb of *Ilaria* has several features similar to modern wombats, particularly in the manus. The shapes of MC II, MC III, the proximal and medial phalanges, and the unguals are nearly identical between the two groups (table 5). The relative length of the phalanges is intermediate between wombats and *Ngapakaldia*. The height of the coronoid process of the ulna, the curve of the scapular blade, and the concave distal surface of the magnum are also features most similar to those in wombats (table 5). On the other hand, the radius is most like that of *Ngapakaldia*, with no notable features except for the thickened distal end common to all vombatoids (table 3). The length of the olecranon process of the ulna is also comparable to that of *Ngapakaldia* (table 4), but is still within the range of digging animals (Hildebrand 1988). The overall form of the scapula and magnum may be peculiar to *Ilaria*, but recall that there is no scapula of *Ngapakaldia* for comparison. The reduced coracoid process and enlarged triceps notch of the scapula are probable synapormorphies for the vombatoids (table 3), and indicate a terrestrial habitus. The retention of a relatively long MC I is an apparently plesiomorphic feature.

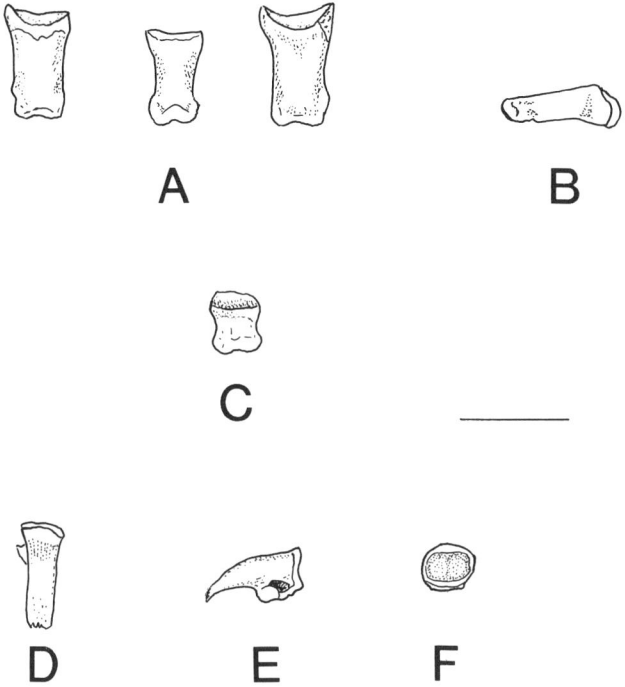

Fig. 27. *I. illumidens*, phalanges, AMNH 102636. Scale = 2 cm.

THE PES (fig. 28)

One complete right fourth metatarsal, one proximal phalanx I, and two fragments of metatarsals are the only elements from *Ilaria*'s hind limbs. The complete metatarsal (AMNH 102638) is most similar to that of the wombats in general shape, especially in the mediolaterally tapered proximal end (table 5 and fig. 28A). The lateral side of the shaft is longer than the medial side, indicating that, as in wombats, this digit turned strongly medially. The distal condyle is not as dorsoventrally flattened as in wombats, nor as much as those of *Ilaria*'s metacarpals, and in this aspect it has a greater resemblance to *Ngapakaldia*.

The two fragments are apparently from the same individual, based on the specimen number, AMNH 102636. The larger fragment is most likely a fourth metatarsal; it consists only of the distal end, equal in size to that of the complete MT IV, and part of the shaft on one side. There is a possibility that it could be a third metatarsal, if like wombats *Ilaria* has secondarily enlarged syndactylous digits. However, the other metatarsal fragment, which consists only of the ventral half of a

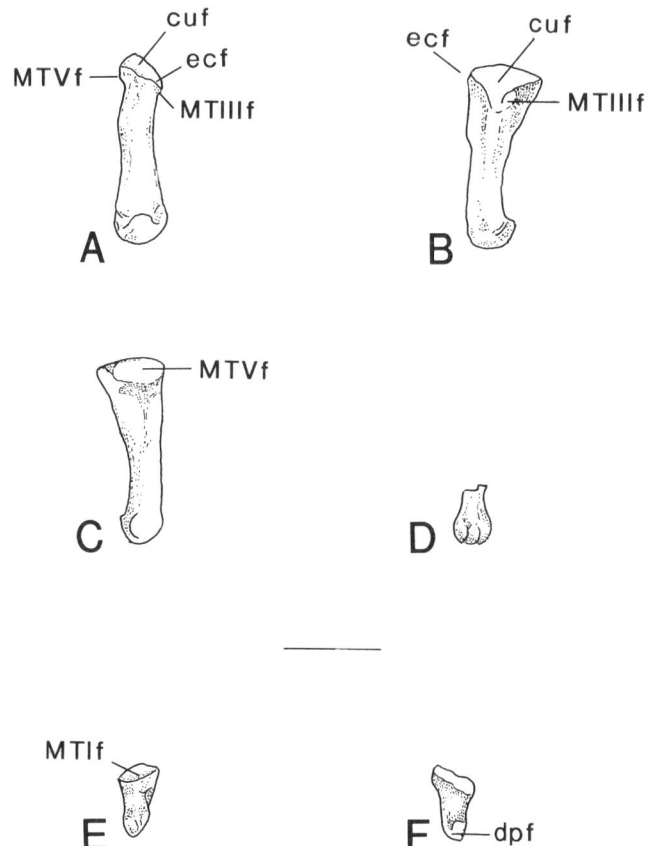

Fig. 28. *I. illumidens*, elements of pes: A. dorsal view, B. medial view, and C. lateral view of right MT IV, AMNH 102638; D. fragment, MT II or III, AMNH 102636, ventral view; E. dorsal view, and F. ventral view of left proximal phalanx I, AMNH 102636. Scale = 2 cm. Abbreviations: cuf, cuboid facet; dpf, distal phalangeal facet; ecf, ectocuneiform facet; f, facet; MT, metatarsal.

distal condyle, is so small that, if indeed it is from the same animal, it could only be from a typically syndactylous digit (fig. 28B), thus ensuring that the other fragment is a MT IV.

The left proximal phalanx I of the pes is nearly complete (some slight erosion on the medial side) and identical with that of *Vombatus*. It has the same strongly tapered, almost conical shape with a rounded distal tip (fig. 28C). This strongly suggests reduction of the first digit as in wombats (table 5). Modern wombats have lost the second phalanx of this digit, and therefore have no distal facet on the first. *Ilaria* does have a tiny distal facet, and therefore may still have had two phalanges in digit I, although the second may only have been rudimentary.

DISCUSSION

At first glance, it appears that *Ilaria* (Aboriginal for "strange") has a structure consistent with its name. It has: a vertebral column most similar to wombats anteriorly and to *Thylacoleo* posteriorly; a forearm intermediate between wombats and palorchestids; and a manus and pes nearly identical with those of wombats, but with a pollex and syndactylous metatarsals like those of *Trichosurus*. The magnum and scapula are unique, but together with the ulna they indicate an animal that carried substantial weight and could generate a great deal of power in the forelimbs. With toes and metacarpals like a wombat, and the high coronoid and long olecranon processes of the ulna, it must be assumed that *Ilaria* was fossorial to some degree (inference based on adaptations for fossorial animals; Shimer 1903, Hildebrand 1988). But because the magnum is not as broad as in wombats and lacks the strong articulation with the unciform, and because the digits are not quite as short, nor the olecranon quite as long, it seems probable that *Ilaria* could not have been as specialized a digger as a modern wombat.

It is especially questionable whether *Ilaria* used its hind feet to aid much in digging. Modern wombats have enlarged the second and third digits of the pes to the point where they no longer appear syndactylous. In fact, the second, third, and fourth digits are enclosed in a common web of skin, and are served by common tendons (Sonntag 1923). This makes the wombat hind foot an effective shovel for kicking out dirt when burrowing (Shimer 1903). If *Ilaria* has typically syndactylous digits, like those of possums and other vombatiforms, it probably could not have been an effective burrower.

Although the evidence for typical syndactyly rests only on one referred fragment of bone, the size and morphology of *Ilaria*'s vertebrae also argue against making too close a functional analogy with the modern wombats. *I. illumidens* is big, clearly larger than *Ngapakaldia tedfordi* (which was about the size of a sheep) and possibly as large or larger than *N. bonythoni* (calf-size). *I. lawsoni*, the other species of *Ilaria* which has no postcranials, is apparently an even larger animal than *I. illumidens*, based on the size of the molar teeth. It is difficult to imagine animals of this size making a burrow. The shape of the sacral vertebra also makes a burrowing habit unlikely. *Ilaria*'s sacrum appears to be narrow transversely and deep dorsoventrally, whereas that of the burrowing wombats is broad and shallow. The short, deep centrum of the lumbar vertebra also indicate a stockier, less flexible lumbar region in *Ilaria* than in wombats. Because of this, it seems unlikely that *Ilaria* could have achieved the same powerful kicking action that modern wombats use in burrowing, or have had the necessary flexibility for maneuvering in a tunnel. A more plausible hypothesis might be an animal that digs for roots, such as a pig.

The similarity between *Ilaria*'s more posterior vertebrae (the ?eighth thoracic, the lumbar and sacral) and those of *Thylacoleo* are of dubious phylogenetic importance. Vertebrae seem to be less reliable bones from which to infer relationships. Most likely, this similarity simply represents convergence in large animals which both had short lumbar regions and narrow sacra, and therefore lacked the flexibility of typical burrowing or arboreal animals. It is interesting to speculate that both animals may have required a strong sturdy lower back to brace against powerful action of the forelimbs: *Ilaria* for scratch-digging, *Thylacoleo* for immobilizing its prey with its club-like forearm and manus (Finch 1971; J.A. Case, pers. comm., 1989).

For deducing relationships, the elements of the manus seem far more reliable. The magnum and metapodials are more complete, unlike the limb bones and verte- brae. More important, the similarities seen in these elements are manifested in the intricate and detailed structure of the facets rather than in general similarities of overall shape. Whereas it could conceivably be convergence that would give *Ilaria* spatulate and rugose metacarpals similar to those of a wombat, the probability of two unrelated animals having the same exact pattern of proximal facets on three separate metapodials (MC II, MC III, MT IV) and the same pattern of phalanges, especially in digit I of the pes, is extremely unlikely. It therefore seems logical to conclude that ilariids are most closely related to vombatids, despite retention of some plesiomorphic features such as typical syndactylous digits and a large MC I.

PHYLOGENETICS

In order to assess the position of *Ilaria* and *Ngapakaldia* in the Suborder Vombatiformes, a character analysis was made of the postcrania of all known species of
vombatiforms and selected species of numerous taxa (dasyurids, bandicoots, possums, and microbiotheriids) representing outgroups (see materials, Appendix I). As
a starting point, the classifications of Woodburne (1984) and Aplin and Archer
(1987) were used to define the Vombatiformes and to identify the outgroup taxa.
The results of the analysis were then used to evaluate these classification schemes,
and more specifically the relationships within the Vombatiformes as proposed by
Marshall, Case, and Woodburne (1990). As Aplin and Archer did not attempt to
show interfamilial relationships in their phylogeny, only their higher-order groupings
(order, infraorder) are subject to evaluation. In their classification scheme, the
Infraorder Vombatomorphia replaces the Superfamily Vombatoidea of Woodburne,
but otherwise they are essentially identical.

Postcranial vombatiform synapomorphies were identified by comparison with the
sister taxon, the Phalangeriformes. Comparison was made primarily with *Trichosurus vulpecula* (the brush-tailed possum) because of its semi-terrestrial habitus.
Assuming that arboreality is plesiomorphic for australidelphian marsupials (Huxley
1880, Bensley 1903, Szalay 1984), the form and habitus of *Trichosurus* make it the
closest living analog of a possible vombatiform ancestor. Two species of *Phalanger*
(*P. ursinus* and *P. vestitus*) and one unidentified species of *Pseudocheirus* were also
examined in order to distinguish apomorphies of *Trichosurus* from true
plesiomorphic arboreal character states.

The validity of apparent synapomorphies was assessed by comparison with the
higher-level outgroups (Peramelina, Dasyuromorphia, and Microbiotheria) and,
when possible, by examination of juveniles and individuals of both sexes (see
materials, Appendix I). Unfortunately, it was difficult to obtain multiple repre-
sentatives of any one species, especially for the bones of the manus and pes, which
are frequently removed in skinning. Still, examination of those specimens available

65

Table 7. Homoplasies of vombatids and *Diprotodon*

1. Cervical vertebrae greatly reduced in length, so that the combined length of C_3-C_7 f< the width of C_7.
2. Broad transverse processes on the lumbar vertebrae.
3. A reduced xiphoid process of the sternum.
4. A large flaring overhang on the deltoid ridge of the humerus.
5. A reduced hamulus on the unciform.
6. A flat or convex (but not concave) MC I facet on the trapezium (decreased mobility of pollex).
7. A laterally expanded proximal phalanx V on both manus and pes.
8. A convex, rather than saddle-shaped, MT I facet on the entocuneiform (hallux not opposable).
9. Relatively short and stout MT II and III (the syndactylous digits).

---

helped to eliminate a number of characters in which the states were too variable to establish clear polarities.

Polarity of the remaining states, which appeared to reflect valid, consistent differences among the various taxa examined, was determined by comparison with all the outgroups mentioned above, but primarily with the arboreal phalangeriforms and the microbiotheriid *Dromiciops*, because of the assumed primitiveness of arboreality. The scarcity of postcrania in the fossil marsupial record made it necessary to rely entirely on modern species for outgroup comparisons, keeping in mind the dangers of assuming the primitiveness of common character states among modern taxa (Szalay 1984). The identification of nearly 90 unique apomorphies for the families of the Vombatiformes further reduced the character list by more than half.

The remaining characters (see tables 2-7 and Appendices II & III) were then analyzed both manually and by computer algorithm to build parsimonious phylogenetic trees. In the manual analysis, and in assessing the computerized results, some character states were given greater confidence (or weight) than others (marked by an asterisk (*) in the tables). Those states that were judged to be more reliable synapomorphies were usually those that involved the detailed shape of metapodial facets, carpals, and tarsals, as opposed to more malleable characters such as proportions of limbs or vertebrae. An exception was the consistent relationship observed between the relative lengths of limb bones (i.e., humerus:radius and femur:tibia). In general, unique complex patterns, especially those that reflect specific functional adaptations, were favored as being less likely to be the result of homoplasy (Szalay 1981).

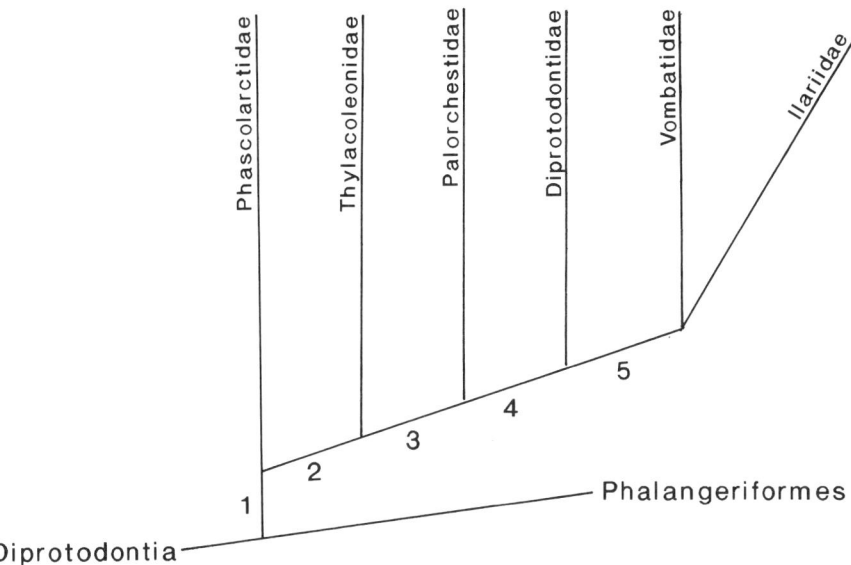

Fig. 29. Postcranial apomorphies of the Vombatiformes (see tables 2-6 for greater detail). Numbers correspond to nodes on the cladogram: 1. Terrestrial modifications to the ankle joint (nos. 6, 7, 8, and 9 in table 2); length of tibia < femur; decreased length of phalanges; increased width, distal end of humerus. 2. Increased width of fifth metapodials (nos. 8, 9, and 12 in table 3); length of radius < humerus; increased stoutness of radius, ulna, and femur (nos. 3, 5, 6, 7, and 10 in table 3); reduced coracoid process on scapula. 3. Increased stoutness of digits, especially MC IV and MT V; digital formula of 3.2.4.5.2 in manus, 4.3.2.5.1 in pes; increased length of olecranon process, ulna; increased width, proximal end of fibula; increased width, iliac crest. 4. Reduction of phalanges, digit I of pes; no distal condyles on MT I; unguals of manus dorsoventrally flattened; distal condyles of MC's flattened; increased thickness at coronoid process of ulna. 5. Expanded and concave proximal end of MC II; broad, convex proximal end of MC III; tapered proximal end of MT IV; reduction of MT I; distal condyles of MT's flattened.

The computer algorithm used was the Wagner analysis of the MIX option in J. Felsinstein's phylogenetic inference package, PHYLIP 2.8 (Felsinstein, 1987). The PHYLIP program consistently generated the same parsimonious tree, even when the order of the taxa was shuffled or data from individual taxa removed.

The results of the MIX analysis are shown in fig. 29. The placement of the outgroups and the position of *Phascolarctos* support the monophyly of the Vombatiformes and Vombatoidea (Vombatomorphia) as defined by Woodburne (1984) and Aplin and Archer (1987). The koalas, despite their superficial resemblance to the arboreal phalangeriforms, are consistently grouped with the other vombatiforms based on the characters examined, especially the shape of the astragalus and the upper ankle joint (see table 2).

The grouping of the vombatoid families differs, however, from the most current phylogeny based on cranial evidence (Marshall, Case, and Woodburne 1990) shown in fig. 1. Both cladograms separate the thylacoleonids from all other vombatoid families and both schemes identify vombatids as a sister group to the diprotodontids, but the positions of the ilariids and palorchestids are quite different. The cranial data recognizes the primitive nature of *Ilaria*'s molar crowns (incipient lophodonty on a basically tribosphenic pattern) compared to the partially lophodont wynyardiids (lower molars only) and the fully lophodont palorchestids and diprotodontids. The origin of the vombatid molar pattern is masked by both a simplified crown pattern and their highly derived hypsodont dual pillar-like form. They are placed closer to the palorchestids and diprotodontids in fig. 1 based on the following apparent synapomorphies: (1) loss of the upper canine, (2) lowering of the inclination angle of the lower incisors, and (3) increased ankylosis of the dentary symphysis.

The postcranial data places ilariids and vombatids most closely to each other, based entirely on the similar structure of the manus and ulna (see table 5). Furthermore, these characters are seen as being the most derived when compared to the postcrania of the other vombatiforms. The palorchestids, with their plesiomorphic possum-like postcranial skeleton, appear as the sister-group to both of these fossorial families as well as to the diprotodontids (represented by the highly derived graviportal *Diprotodon*). Due to a lack of sufficient postcranial characters, the wynyardiids were not considered in this analysis.

In attempting to reconcile these two conflicting phylogenetic schemes, one must examine the characters responsible for the differences to determine which seem most valid, and construct a tree that reflects the strongest points of both systems. The primary basis for the placement of the ilariids, wynyardiids, palorchestids, and diprotodontids in the Marshall, Case, and Woodburne scheme (fig. 1) is the trans-formational sequence of increasing lophodonty, which views the ilariids as most primitive and palorchestids and diprotodontids as the most derived. To rearrange this pattern as in the postcranial tree (fig. 29), would necessitate fully lophodont molars developing separately in the palorchestids and diprotodontids, or being reversed (lost) in the ilariids. Since these possibilities are neither logical nor parsimonious, it seems best to leave these families alone, in the pattern shown in fig. 1.

The position of the vombatids, on the other hand, may be more flexible. As mentioned above, the only cranial characters that group them with the dipro-todontids and palorchestids are the loss of the upper canine, the decreased angle of the lower incisors, and an increased length of the dentary symphysis. The highly derived, hypselodont, rodent or lagomorph-like molars mask their own origin; they may or may not have been derived from a more lophodont tooth than seen in the ilariids. In fact, the unworn molar crowns of *Vombatus*' pouch young show a

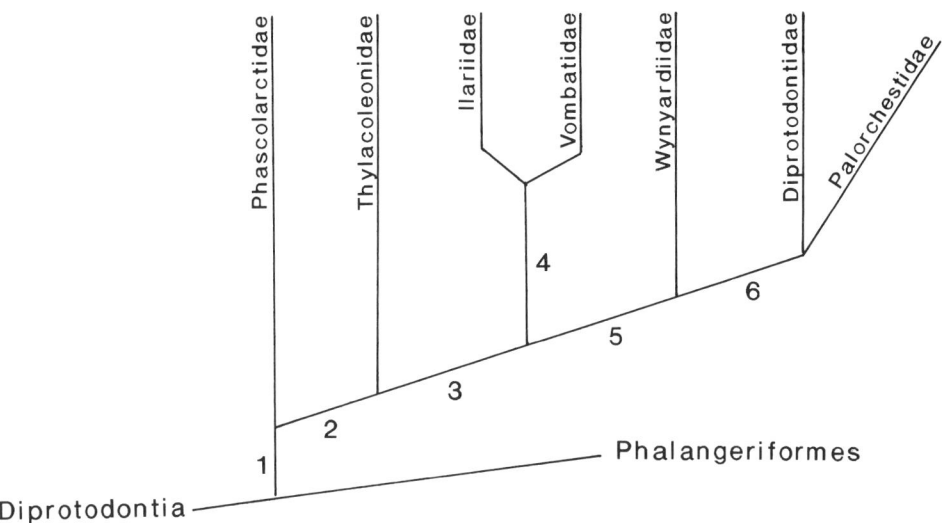

Fig. 30. Major skeletal apomorphies of the Vombatiformes synthesized from figs. 1 and 29 (cranial apomorphies after Marshall, Case, and Woodburne, 1990): 1. Reduction of functional premolars to P3 above and below; terrestrial modifications to the ankle joint. 2. Bilaminar bulla; increased stoutness of limbs and fifth metapodials. 3. Squamosal bulla; increased stoutness of digits; decreased length of MC IV and MT V. 4. Incipient lophodonty; reduction of digit I of pes; dorsoventrally flattened metapodials and unguals; fossorial habitus. 5. Trend toward lower molar bilophodonty. 6. Molars completely lophodont; pelvis deep; epipubics thick; reduced radial notch on ulna.

somewhat selenodont pattern (Archer 1984a). Postcranially, the similarity between the metapodials and phalanges of the vombatids and ilariids seems too great to be homoplasic (see figs. 9 and 10). While fossorial adaptations of the manus could certainly have developed twice in the Vombatoidea, it is extremely unlikely that both familes would have developed the same unique shape of the proximal facets seen in both the metacarpals (figs. 9 and 10) and metatarsals. It therefore seems logical to place the vombatids with the ilariids in a common fossorial clade.

While the postcranial evidence for keeping the ilariids and vombatids together appears strong, that for grouping them with the diprotodontids is not. None of the similarities between vombatids and diprotodontids (table 7) are sufficiently complex that they could not be the result of homoplasy. In fact, many of the similarities of the manus and pes (i.e., flattened distal condyles of the metapodials, laterally expanded proximal phalanx V) are simply features that one would expect to see in both graviportal and fossorial animals.

The most parsimonious phyletic interpretation, then, that incorporates the most probable synapomorphies of both the cranial and postcranial data, is that seen in

Table 8. Postcranial synapomorphies of Diprotodontians

		DIPROTODONTIANS	Other AUSTRALIDELPHIANS
*1.	Scapholunar	Fused	Scaphoid and lunate separate
*2.	Cuboid	Contacts astragalus	No contact
3.	Shaft of humerus	Straight	Curves anteroposteriorly

fig. 30. It is identical with that of Marshall, Case, and Woodburne (fig.1) except for the position of the vombatids. Allying the vombatids down with the ilariids requires the assumption that the postcranial synapomorphies linking the vombatids and ilariids have greater weight than the cranial character states that link the vombatids to the diprotodontids and palorchestids. This implies that: (1) vombatid molars are derived from a more plesiomorphic, ilariid-like dentition; (2) vombatids lost the upper canine, decreased the angle of the lower incisors, and increased the length of the dentary symphysis independent of the palorchestids and diprotodontids; (3) the unique fossorial manus and pes of the vombatids and ilariids originated only once; (4) palorchestids retained a relatively plesiomorphic postcranial skeleton while developing a more derived lophodont dentition; (5) the lophodont molars of the palorchestids and diprotondontids are the end result of a transformational sequence beginning with the incipient lophodonty that probably existed in the common ancestor of the ilariids and wynyardiids; and (6) similarities in the postcrania between wombats and *Diprotodon* are the result of symplesiomorphies and homoplasy.

The postcranial synapomorphies for each node on the phyletic tree in fig. 30 (and fig. 29) are explained in greater detail in tables 2-6. In addition, table 8 shows postcranial synapomorphies for the order Diprotodontia, which is comprised of the two sister suborders Vombatiformes and Phalangeriformes. Character states that appear to be more reliable, and were therefore given greater weight in the analysis, are marked with an asterisk (*).

POSTCRANIAL EVOLUTION
IN THE VOMBATIFORMES

SYNAPOMORPHIES OF THE VOMBATIFORMES

An analysis of the synapomorphies at each node of the cladogram in fig. 30 (tables 2-6) shows trends in the functional adaptations of the vombatiform postcranial skeleton. In the synapomorphies of the vombatiforms (see table 2), we see adaptations for slightly increased size and terrestrial locomotion. The phalanges are relatively shorter, MC I is stouter, and the distal humerus broader than in the arboreal phalangeriforms. Most distinctive are the changes in the ankle joint, especially the concave dorsal surface of the astragalus with its medial tibial knob and a distinct, but reduced, lateral fibular facet. This arrangement begins to approach the tenon-mortise structure described by Szalay (1984) as being typical of a terrestrial habitus, since it confers greater stability to the ankle joint. In the arboreal phalangeriforms and *Dromiciops* (Microbiotheria), the dorsal surface of the astragalus is smoothly convex, with no distinct division between the tibial and fibular facets, and the fibular facet is nearly as large as the tibial. The anterior tibial crest is oriented between the first and second digits so that digits II through V extend laterally.

It is notable that the koalas, while sharing many features with the arboreal possums, also show these modifications to the astragalus (although to a lesser degree) seen in the other, fully terrestrial, vombatiforms (fig. 31). This helps support the hypothesis that koalas have returned to the trees from a terrestrial ancestry. This suggests that koalas may have a less flexible upper ankle joint than possums, a proposition which is supported by observations of how they climb. Instead of branch-walking with laterally extending toes of the pes and a prehensile tail, modern koalas climb more like bears, clinging to branches with powerful adductor muscles of the thigh. In fact, one of their most distinctive postcranial apomorphies is the greatly shortened ischium, which would confer greater mechanical advantage for adduction rather than extension of the leg by the adductor muscles (Elftman

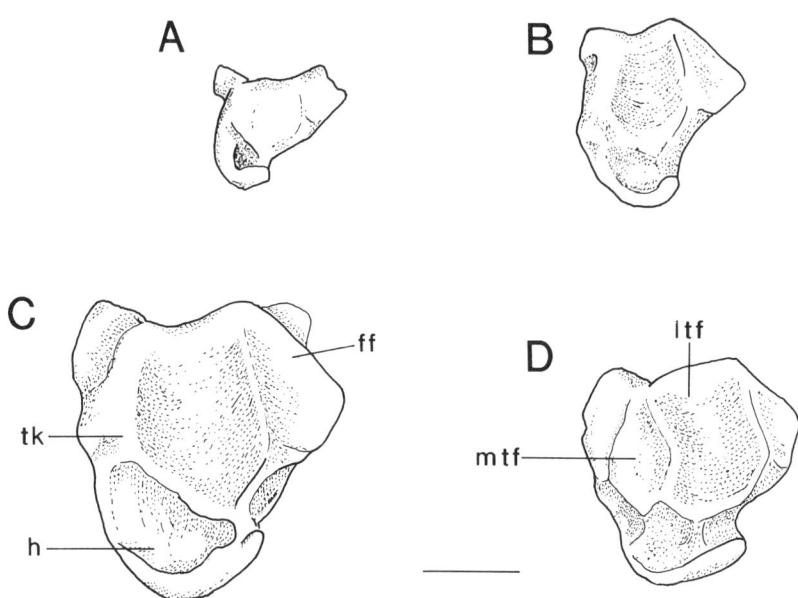

Fig. 31. Dorsal left astragali of A. *Trichosurus vulpecula*; B. *Phascolarctos cinereus*; C. *Ngapakaldia tedfordi*; D. *Vombatus ursinus*. Although *Phascolarctos* and *Trichosurus* are both arboreal animals, note the greater similarity of *Phascolarctos'* astragalus to the two terrestrial Vombatiformes, *Ngapakaldia* and *Vombatus*. Scale = 1 cm. Abbreviations: h, head; ff, fibular facet; ltf, lateral tibial facet; mtf, medial tibial facet; tk, tibial knob.

1929). Other postcranial apomorphies of koalas include the forcipate manus (opposable first and second digits) and the loss of two pairs of ribs.

SYNAPOMORPHIES OF THE VOMBATOIDEA

The adaptations of the vombatoids (table 3) can all be related to increased size, complete terrestriality, and more weight borne on the front limbs. The changes in the limbs and feet (characters 1, 2, 5, and 6 in table 3) indicate a four-footed posture with the weight borne primarily on the lateral digits of the pes (character 12), and showing increased use of the fifth digit of the manus as well (characters 7 and 8). As cursorial animals typically lengthen the distal limb segments (Hildebrand 1988), the shortened radius (character 3) indicates a common noncursorial habitus for this group. Since all of the vombatoids observed are large animals, there is a possibility that many of these character states reflect convergence for increased size, rather than being synapomorphies. However, previous classification schemes (e.g., Aplin and Archer 1987; Marshall, Case, and Woodburne 1990) have also grouped these families together based on cranial and dental characteristics, so it is more likely that being a large terrestrial animal is a synapomorphy of the vombatoids.

DIVERSITY WITHIN THE VOMBATOIDEA

The thylacoleonids (fig. 30) maintained the plesiomorphic vertebral formula (13 thoracic and 6 lumbar vertebrae), including a long tail (Finch and Freedman 1986) and an apparently plantigrade pes (Wells and Nichols 1977). The most distinctive apomorphic feature of *Thylacoleo* (besides the cranial adaptations for carnivory) is the digitigrade manus with its uniquely enlarged and flexible first ungual. The medial Miocene thylacoleonid *Priscileo* is represented postcranially by only a humerus, a radius, an ectocuneiform, and a magnum (Rauscher 1987). The structure of the humerus and radius is consistent with the synapomorphies of the vombatoids (table 3), but little else can be inferred from this material, or from the highly derived *Thylacoleo*, regarding the relationship of this family to the other vombatiforms.

In table 4, we see character states that unite the ilariids, vombatids, palorchestids, and diprotodontids, but exclude the thylacoleonids. Again, the same possibility of convergence presents itself, since all the species dealt with are large animals and most of the shared character states (even the most reliable ones) are not complex features, but rather bone proportions associated with increased size. The adaptations for increased strength of the forelimb, for example, could have developed separately in the diprotodontids and palorchestids, families which show a trend toward large graviportal animals, and in the vombatids and ilariids, families which show a trend toward a fossorial manus. Both lifestyles call for stout, strong bones as seen in the ulna and manus, and none of the features involved are sufficiently uniquely complex to indicate without doubt a sister relationship between these two groups. Cranial and dental morphology indicates a close relationship between these taxa and with wynyardiids (Marshall, Case, and Woodburne 1990), but the postcranial morphology points only to a similar build and intriguing possibilities of relationship.

Although the only major synapomorphies between ilariids and vombatids (table 5) are in the forearm and feet, they are all very complex, uniquely similar features. The proximal and distal ends of the metapodials and phalanges are remarkably similar and distinct from the plesiomorphic pattern seen in the arboreal phalangers and early palorchestids (figs. 9 and 10). Both families show adaptations in the forelimbs for a fossorial lifestyle, and an identical reduction of the hallux in the pes.

The shapes of the scapula, radius, and vertebrae (except for the atlas), however, are distinct between the two families and contain no important synapomorphies.

The ilariids diverged from the vombatids by becoming large browsing animals, while the dentition of the vombatids became more rodent-like. Modern wombats have re-enlarged the second and third digits of the pes, creating a shovel-like pes as well as manus. Other adaptations for burrowing include the flaring ilium, long olecranon process, and dorsoventrally flattened centra of the vertebrae. The tibial crest twists medially to shift the weight of the hind foot over the re-enlarged syndactylous digits. Unfortunately, we have no pelvis or tibia of *Ilaria* to compare with those of wombats, nor were syndactylous digits confidently identified.

The postcrania of the wynyardiids are, as yet, little known, but the position of this family on the cladogram (fig. 30) is based on the dental characteristics that place them between ilariids and palorchestids (Marshall, Case, and Woodburne 1990). A tantalizing glimpse at the pes of *Muramura williamsoni* (N.S. Pledge, pers. comm., 1989), a medial Miocene wynyardiid, shows a structure apparently similar to that of *Ngapakaldia*, but small, more like a wombat in size. It is also described as having a short tail (Pledge 1987), an apparently derived feature, seen so far only in wombats and koalas.

The synapomorphies between the diprotodontids and the palorchestids (table 6) are extremely few and of poor quality. This is due, in part, to having only the highly derived, graviportal, Pleistocene *Diprotodon* available for comparison with the medial Miocene palorchestids. But it also seems that, aside from graviportal specializations in later diprotodontids, the postcranial skeleton in both these families is essentially plesiomorphic, and there may be no synapomorphic character states other than those in table 6. Waters (1967) described the postcrania of a number of late Miocene (Alcoota fauna) diprotodontids, with the observation that they were frequently intermediate in structure between that of *Ngapakaldia* and *Diprotodon*.

The graviportal adaptations of the diprotodontids appear to be superimposed upon the plesiomorphic pattern of the early palorchestids. They have the same vertebral formula and long tail, and what little postcrania are available from Miocene diprotodontids are similar to those of *Ngapakaldia* (Waters 1967). Few postcrania are available from the Pliocene-Pleistocene palorchestid, *Palorchestes azael*. Only the forelimbs and caudal vertebrae have been described (Bartholomai 1962, Archer and Bartholomai 1978), and it is not known at this time whether or not the plesiomorphic pattern of the manus and pes is still retained in this later genus.

Tables 7 and 9 list homoplasies observed in the postcrania of wombats (including the giant Pleistocene wombat *Phascolonus*), *Diprotodon*, and *Phascolarctos*. They are not considered symplesiomorphies of vombatiforms because of their absence in argue in favor of a shared derivation. They are interesting similarities, but in my

Table 9. Homoplasies of vombatids, *Diprotodon*, and *Phascolarctos*

1. Reduction of the ventral flange on the transverse process of the sixth cervical vertebra.
2. A broad, stocky manubrium of the sternum.
3. A dorsoventrally flattened ilium (less so in *Phascolarctos* than the other two).
4. An expanded navicular facet on the entocuneiform.
5. Broad, flattened, distal condyles on the metapodials.

---

opinion do not outweigh the synapomorphies expressed by the cladogram in fig. 30.

Most of these features indicate only a similarity of body shape and the similar effects of large body weight or fossorial activity (both, in the case of *Phascolonus*) on lateral expansion of the digits. However, two unexplained problems remain: the similar shape of C_6, and the expanded distal condyles of the metapodials in the arboreal koala. Again, postcrania of earlier members of each family are needed to understand the development of these features, but they are probably coincidental.

CONCLUSION

The postcrania of *Ngapakaldia tedfordi* and *Ilaria illumidens* shed interesting light on the skeletal phylogeny of the Vombatiformes. *Ngapakaldia*'s general similarity in form to *Trichosurus* (especially in the vertebrae, manus, and pes) reflects the arboreal ancestry of the vombatiform clade, and indicates that this possum-like structure was the plesiomorphic state from which the more specialized vombatoid families (fossorial wombats and ilariids, digitigrade thylacoleonids, and graviportal diprotodontids) are derived. The terrestrial adaptations of *Ngapakaldia* that are superimposed upon this primitive pattern indicate an unspecialized sheep-size plantigrade animal with fairly strong forelimbs. The highly derived, lophodont teeth (Stirton 1967) round out the picture of this browser.

Ilaria is intriguingly opposite to *Ngapakaldia* in that it has a more plesiomorphic vombatiform dentition (Tedford and Woodburne 1987; Marshall, Case, and Woodburne 1990), but more derived postcrania. With distinctly similar metapodials and phalanges, ilariids and vombatids appear to form a clade with adaptations for some degree of fossorial activity. The size of *I. illumidens*, the even larger *I. lawsoni* (Tedford and Woodburne 1987), and the morphology of the vertebrae (deep centra, narrow sacrum) seem to indicate a much lesser degree of digging ability than in the burrowing wombats. Still, one can envision this calf-sized browser using its spatulate digits to scratch-dig for roots or tubers.

A phyletic analysis made of the postcrania of the vombatiform families and several species of outgroups turned out to disagree with the most recent work done on this group using cranial data (Marshall, Case and Woodburne 1990). The positions of the phascolarctids and thylacoleonids are in agreement, but the derived skeleton and primitive dentition of the ilariids are at odds with the derived dentition and plesiomorphic skeleton of the palorchestids. The most probable solution appears to be the sister-group relationship of vombatids and ilariids. This rests on the assumption that the rodent-like hypselodont cheek teeth of wombats must be derived not from a lophodont ancestor, but from a more primitive ilariid-like dentition. This solution results from placing greater confidence in weighting *a priori* (a) the transformational sequence of increasing lophodonty seen in the vombatiform families according to the Marshall, Case, and Woodburne scheme (fig. 1) and (b) the similarity of the manus between ilariids and vombatids over other cranial or post-cranial features. This, however, implies that all similarities between wombats and *Diprotodon* (both cranial and postcranial) are the result of symplesiomorphy or homoplasy.

The comparative study of known vombatiform postcrania, as organized in the phylogenetic tree in fig. 30, revealed some interesting trends. All vombatiforms, including the phascolarctids, appear to have had a terrestrial ancestor. Probable evidence for this exists in the similar shape of the koala astragalus to those of the other, fully terrestrial vombatoids, especially in its concave dorsal surface and more laterally facing fibular facet. The vombatoids (the remaining vombatiform families exclusive of the phascolarctids) all appear to be fully terrestrial, plantigrade animals. They can be further divided into three clades: the carnivorous thylacoleonids, with a digitigrade manus in the Pleistocene *Thylacoleo*; the fossorial ilariids and vombatids; and the wynyardiids, palorchestids, and diprotodontids, who show a derived lophodont dentition but have kept a basically plesiomorphic postcranium. Although the diprotodontids have graviportal adaptations and show several homoplasies in common with the vombatids, their lophodont teeth and several plesiomorphic features of their postcranium indicate their close relationship to the palorchestids.

That both *Ngapakaldia* and *Ilaria* are of medial Miocene age shows that the split between the generalized plantigrade animals and those with fossorial adaptations took place well before this time. In fact, all of the nodes on fig. 30 must have predated the medial Miocene. Work by Case (1989) and Springer (1988) indicates that the time of vombatiform radiation was probably during the Oligocene, based on changes in vegetation and DNA molecular clock data.

The lack of Australian terrestrial deposits of Eocene or Oligocene age leaves the story of vombatiform origins rather incomplete at this time, a problem common to all of the Australidelphian orders. But work continues, in the search for paleogene deposits as well as in studies of neogene specimens. Current studies of early thylacoleonids (R.T. Wells, pers. comm., 1989), the wynyardiid *Muramura* (N.S. Pledge, pers. comm., 1989), and an unknown vombatoid (vombatoid genus A, Tedford et al. 1977) will hopefully shed new light on the relationships of this highly diversified marsupial suborder.

Appendix I

List of specimens used for study and in the phylogenetic analysis. Fossil material is denoted by an asterisk (*), and all specimens observed are adults unless otherwise noted.

Abbreviations used in the specimen list stand for the following institutions:

AMNH American Museum of Natural History, New York
LACM   Los Angeles County Museum of Natural History
MVZ   Museum of Vertebrate Zoology, University of California, Berkeley
UCMP Museum of Paleontology, University of California, Berkeley
UCR   University of California, Riverside

Microbiotheria
    Microbiotheridae
        Dromiciops australis, MVZ 163431; UCR
Dasyuromorphia
    Dasyuridae
        Dasyuroides byrnei, UCR
        Dasyurus maculatus, MVZ 127003, male
        Sminthopsis murina, UCR
Peramelina (bandicoots)
    Peramelidae
        Echymipera kaluba, MVZ 132750, male; MVZ 138478, male
        Perameles gunni, MVZ 133133, female
        Isoodon macrourus, MVZ 140130, juvenile female
        Isoodon obesulus, MVZ 129753, juvenile male

Thylacomyidae
  Macrotis lagotis, Greater Bilby, UCR
Diprotodontia
  Phalangeriformes (possums)
    Pseudocheiridae
    Phalangeridae
      Phalanger ursinus, MVZ 125528, female
      Phalanger vestitus, MVZ 132753, male
      Trichosurus vulpecula, MVZ 127142, male;
        MVZ 127148, male; UCR, juvenile
  Vombatiformes
    Phascolarctidae (koalas)
      Phascolarctos cinereus, LACM MG73; LACM 640, male;
        LACM 641, female; MVZ 119477, juvenile; MVZ 121453, male
    Vombatidae (wombats)
      Vombatus ursinus, LACM 54088, juvenile; MVZ 133203, female;
        MVZ 133287, juvenile male; UCR
        (MVZ specimens of *V. ursinus* are labeled as *V. mitchelli*)
      Lasiorhinus latifrons, MVZ 122097, male;
        MVZ 122098, juvenile female; MVZ 122099, female;
        MVZ 125937, male; MVZ 129754, juvenile
     Phascolonus gigas (Pleistocene), descriptions from Stirling (1913) and
        Rich, Van Tets, and Knight (1985)
    *Thylacoleonidae
      Priscileo pitikatensis, UCMP 126577 (medial Miocene), and descriptions
        by Rauscher (1987)
      Thylacoleo carnifex, P12384 (Pleistocene) from the South Australian
        Museum, and the Moree specimen (Pleistocene) from the Australian
        Museum, Sydney, plus descriptions by Owen (1883a,b), Wells and
        Nichols (1977), Finch (1982), and Finch and Freedman (1986)
     *Wynyardiidae
      Wynyardia bassiana, PM 4421 (early Miocene) from the Chicago Natural
        History Museum; #40821 (another cast of the same specimen) from
        the University of Texas; plus a description by Ride (1964)
    *Ilariidae
      Ilaria illumidens, AMNH 102185, 102221, 102636, 102638, 102644,
        102649 (holotype), and 102651 (medial Miocene)
      Vombatoid genus A, AMNH 102646 (medial Miocene)
      (Reference for both specimens: Tedford and Woodburne 1987)

* Palorchestidae
   Ngapakaldia tedfordi, UCMP 60984, 69812, 69813, 71416, 72130, 126722
       (casts of holotype, SAM P13851) (medial Miocene)
   Ngapakaldia bonythoni, UCMP 57257, 57258 (casts of holotype plus
       original material) (medial Miocene)
   Pitikantia dailyi, UCMP 60981 (medial Miocene)
   (Reference for all 3 species: Stirton 1967)
* Diprotodontidae
   Information on the Pleistocene species, *Diprotodon optatum*, and other
   diprotodontids is taken from Stirling and Zietz (1899) and Waters (1967)

Appendix II

Characters used in the phylogenetic analysis. Numbers correspond to the character distribution in appendix III. (A) denotes a derived character state; (B) denotes a primitive state (1 and 0 in appendix III, respectively). Due to the nature of the program used (PHYLIP 2.8, MIX program), characters with intermediate states have been rewritten as two or more characters with only two possible states (A or B). See "Terminology and Materials" for definitions of length, width, and height (thickness).

1. Cervical vertebrae: combined length of C_3-C_7 divided by width of C_7 (A) < 1.0 (short neck); (B) > 1.0 (long neck).
2. Cervical vertebrae: ventral flange on the transverse process of C_6 (A) reduced; (B) prominent.
3. Lumbar vertebrae: transverse processes (A) extremely broad anteroposteriorly; (B) narrow.
4. Lumbar vertebrae: transverse processes (A) extend perpendicular to sagittal plane; (B) in sagittal plane.
5. Lumbar vertebrae: angle between anterior articular processes and the long axis of the centrum (A) > 50°; (B) < 40°.
6. Sacral vertebra #1: height of centrum divided by width (A) $\geq$ 0.64; (B) < 0.64.
7. Sacral vertebra #1: height of centrum divided by width (A) $\geq$ 0.75; (B) < 0.75.
8. Sternum: xiphoid process (A) reduced; (B) not reduced.
9. Sternum: max. width of manubrium divided by length (A) > 0.5; (B) $\leq$ 0.5.
10. Sternum: max. width of manubrium divided by length (A) $\geq$ 0.5; (B) < 0.5.
11. Scapula: coracoid process (A) reduced; (B) extends medial to glenoid fossa.
12. Scapula: triceps notch (A) large; (B) small.
13. Scapula: medial surface (A) concave; (B) flat.

14. Humerus: shaft (A) straight; (B) curves anteroposteriorly.
15. Humerus: max. width across distal end divided by length (A) > 0.3; (B) < 0.3.
16. Humerus: deltoid ridge (A) has a large flaring overhang; (B) has no lateral overhang.
17. Ulna: anteroposterior thickness at semilunar notch divided by bone length (A) > 0.1; (B) ≤ 0.1.
18. Ulna: thickness at semilunar notch divided by bone length (A) ≥ 0.1; (B) < 0.1.
19. Ulna: length of olecranon process divided by bone length (A) ≥ 0.20; (B) < 0.20.
20. Ulna: length of olecranon process divided by bone length (A) ≥ 0.25; (B) < 0.25.
21. Ulna: radial notch (A) absent; (B) prominent.
22. Ulna: max. anteroposterior thickness at coronoid process divided by bone length (A) > 0.2; (B) < 0.2.
23. Ulna: max. thickness at coronoid process divided by bone length (A) ≥ 0.2; (B) < 0.2.
24. Radius: length (A) ≤ length of humerus; (B) > length of humerus.
25. Radius: anteroposterior thickness of distal end divided by bone length (A) > 0.1; (B) ≤ 0.1.
26. Radius: thickness of distal end divided by bone length (A) ≥ 0.1; (B) < 0.1.
27. Cuneiform of manus: positioned (A) more proximally than the scapholunar; (B) directly lateral to scapholunar.
28. Pisiform: area of ulnar facet (A) ≥ area of cuneiform facet; (B) < cuneiform facet.
29. Pisiform: mediolateral width divided by anteroposterior thickness (A) > 0.5; (B) ≤ 0.5.
30. Pisiform: width divided by thickness (A) ≥ 0.5; (B) < 0.5.
31. Scapholunar: (A) fused; (B) scaphoid and lunate separate.
32. Magnum: MC III facet (A) fully concave; (B) convex or keeled.
33. Unciform: area of MC V facet (A) ≥ area of MC IV facet; (B) < MC IV facet.
34. Unciform: hamate process (A) reduced; (B) large.
35. Trapezium: MC I facet (A) flat or convex; (B) concave or saddle-shaped.
36. All metapodials: distal condyles (A) flattened; (B) spherical.
37. Metacarpal I: max. width divided by length (A) > 0.4; (B) ≤ 0.4.
38. MC I: max. width divided by length (A) > 0.5; (B) ≤ 0.5.
39. MC II: trapezoid facet faces (A) posteriorly; (B) medially.
40. MC II: proximal end is (A) expanded and concave; (B) mediolaterally compressed.

41. MC III: length (A) < 3 times the mid-shaft width; (B) > 3X width.
42. MC III: facet for MC IV is (A) dorsal to ligamental pit (may have a proximal component as well); (B) proximal only to ligamental pit (no dorsal component).
43. MC III: facet for MC IV is (A) dorsal to ligamental pit only (no proximal component); (B) is proximal to ligamental pit (may have a dorsal component).
44. MC III: magnum facet (A) convex; (B) grooved or concave.
45. MC III: outline of magnum facet is (A) semicircular; (B) triangular.
46. MC III: outline of magnum facet is (A) dorsoventrally higher than wide; (B) wider than high, or equally wide as high.
47. MC III: MT II facet is (A) continuous with the magnum facet; (B) separate, distinct, and lateral to the magnum facet (fig. 10).
48. MC IV: (A) reduced, digital formula = 3.2.4.5.1; (B) nearly as long as or longer than MC III.
49. MC IV: proximal end (A) higher dorsoventrally than wide; (B) height = width, or wider than high.
50. MC V: max. width of distal end divided by length (A) $\geq$ 0.5; (B) much less than 0.5.
51. Proximal phalanges of manus or pes: (A) much shorter than corresponding MC or MT; (B) nearly as long as or longer than corresponding MC or MT.
52. Proximal phalanges of manus, IV and V only in pes: mid-shaft width divided by length (A) $\geq$ 0.3; (B) < 0.3.
53. Proximal phalanges of manus, IV and V only in pes: mid-shaft width divided by length (A) $\geq$ 0.4; (B) < 0.37.
54. Proximal phalanges of manus, IV and V only in pes: mid-shaft width divided by length (A) > 0.4; (B) $\leq$ 0.4.
55. Proximal phalanges, manus or pes: distal ends are (A) dorsoventrally tapered, so that the articulation for the medial phalanx faces ventrally; (B) not tapered.
56. Proximal phalanx V of manus or pes: (A) laterally expanded; (B) not laterally expanded.
57. Medial phalanx III of manus or IV of pes: length divided by length of corresponding metapodial (A) < 0.5; (B) > 0.5.
58. Unguals of manus or pes: (A) most are dorsoventrally flattened; (B) most are mediolaterally compressed.
59. Unguals of manus or pes: (A) all are dorsoventrally flattened; (B) some or all may be mediolaterally compressed.
60. Pelvis: dorsoventral height at symphysis divided by width at mid-ischium (A) $\geq$ 0.8; (B) < 0.8.

61. Pelvis: ilium (A) more dorsoventrally flattened than mediolaterally; (B) more mediolaterally flattened.
62. Pelvis: ilium (A) greatly dorsoventrally flattened; (B) only slightly dorso-ventrally, or mediolaterally flattened.
63. Pelvis: width of iliac crest divided by length of ilium (A) $\geq$ 0.5; (B) < 0.5.
64. Pelvis: width of iliac crest divided by length of ilium (A) > 0.5; (B) $\leq$ 0.5.
65. Epipubics: width at mid-bone divided by length X 100 (A) > 10; (B) < 10.
66. Femur: width of proximal end divided by length (A) $\geq$ 0.3; (B) < 0.3.
67. Femur: width of proximal end divided by length (A) 0.3; (B) $\leq$ 0.3.
68. Tibia: length (A) $\leq$ length of femur; (B) > length of femur.
69. Tibia: length (A) < length femur; (B) $\geq$ length of femur.
70. Tibia: (A) ligamental pit posterior to medial malleolus; (B) no pit.
71. Tibia: astragalar facet (A) not saddle-shaped; (B) saddle-shaped.
72. Tibia: when articulated with pes, (A) tibial crest is not aligned between digits I and II (pes turns in medially); (B) is aligned between digits I and II (pes turns out laterally).
73. Fibula: lateral notch on distal end (A) extremely slight or absent; (B) clearly present.
74. Fibula: lateral notch on distal end (A) reduced or absent; (B) deep.
75. Fibula: width of proximal end divided by bone length (A) $\geq$ 0.1; (B) < 0.1.
76. Fibula: width of proximal end divided by bone length (A) > 0.1; (B) $\leq$ 0.1.
77. Fibula: femoral facet (A) present; (B) not present.
78. Fibula: angle between the femoral facet and the flabellar facet (A) $\geq$ 100°; (B) < 100°.
79. Fibula: angle between the femoral facet and the flabellar facet (A) > 100°; (B) $\leq$ 100°.
80. Astragalus: width of fibular facet divided by width of tibial facet (A) < 0.75; (B) > 0.75.
81. Astragalus: lateral tibial facet (A) concave; (B) convex.
82. Astragalus: head (A) narrow, mediolaterally compressed, "wheel-shaped"; (B) wider than long.
83. Cuboid: (A) contacts astragalus; (B) no contact.
84. Cuboid: area of MT V facet (A) $\geq$ area of MT IV facet; (B) < MT IV facet.
85. Ectocuneiform: proximodistal length (A) < dorsoventral height; (B) $\geq$ height.
86. Ectocuneiform: length divided by height (A) < 0.8; (B) $\geq$ 0.8.
87. Entocuneiform: MT I facet (A) convex; (B) saddle-shaped (concavo-convex).
88. Entocuneiform: navicular facet (A) longer than mesocuneiform facet; (B) shorter than or equal to mesocuneiform facet.
89. Digit I of pes: (A) reduced, nonfunctional; (B) large, opposable.

90. Metatarsal I: entocuneiform facet (A) strong concave component; (B) mostly convex.
91. MT I: (A) reduced, does not bear much weight; (B) large, weight-bearing.
92. MT I: distal end (A) convex, no condyles; (B) concave, with condyles.
93. MT II and III: (A) syndactylous; (B) not syndactylous.
94. MC II and III: mid-shaft width divided by bone length (A) $\geq$ 0.15; (B) < 0.15.
95. MT IV: proximal end (A) mediolaterally tapered; (B) square.
96. MT IV and V: (A) either one or both increased in length over MT III; (B) MT III longest, digital formula 3.2.4.5.1.
97. MT V: length (A) secondarily reduced, digital formula 4.3.2.5.1; (B) formula not 4.3.2.5.1.
98. MT V: (A) longest; (B) not longest.
99. MT V: max. width divided by length (A) $\geq$ 0.7; (B) < 0.7.
100. MT V: max. width divided by length (A) > 0.7; (B) $\leq$ 0.7.b

Appendix III

Distribution of states of 100 characters among 11 marsupial families.
0 = plesiomorphic condition; 1 = apomorphic; ? = material missing.

	1	2	3	4	5	6	7	8	9	10	11	12	13	14	15	16	17
Microbiotheridae	0	0	0	0	0	0	0	?	?	?	0	0	0	0	0	0	0
Dasyuridae	0	0	0	0	0	0	0	?	0	?	1	0	0	0	0	0	0
Peramelidae	0	0	0	0	0	0	0	0	0	0	1	1	0	0	0	0	0
Pseudocheiridae	0	0	0	0	0	0	0	0	0	1	0	1	0	1	0	0	0
Phalangeridae	0	0	0	0	0	0	0	0	0	1	0	0	0	1	0	0	0
Phascolarctidae	0	1	0	1	0	1	0	0	1	1	0	0	0	1	1	0	0
Thylacoleonidae	0	0	0	1	1	1	1	?	?	?	1	1	0	1	1	0	0
Palorchestidae	?	?	0	1	0	?	?	?	?	1	?	?	?	1	1	0	1
Diprotodontidae	1	1	1	1	0	1	0	1	1	1	1	1	0	1	1	1	1
Ilariidae	?	?	?	1	1	1	1	?	?	?	1	1	1	?	?	?	1
Vombatidae	1	1	1	1	0	0	0	1	1	1	1	1	1	1	1	1	1

(Appendix III, cont.)

	18	19	20	21	22	23	24	25	26	27	28	29	30	31	32	33	34
Microbiotheridae	0	0	0	0	0	0	0	0	0	0	0	0	0	0	1	0	0
Dasyuridae	0	0	0	0	0	0	0	0	0	0	0	0	1	0	1	1	0
Peramelidae	0	1	0	0	0	0	1	0	0	0	0	0	0	0	0	0	1
Pseudocheiridae	0	0	0	0	0	0	1	1	1	0	0	0	0	1	0	0	0
Phalangeridae	0	0	0	0	0	0	0	0	0	0	0	0	0	1	0	0	0
Phascolarctidae	0	0	0	0	0	0	0	0	0	0	0	0	1	1	0	0	0
Thylacoleonidae	1	0	0	0	0	1	1	1	1	1	0	1	1	0	1	1	0
Palorchestidae	1	1	0	1	0	0	1	0	1	1	1	1	1	0	0	1	0
Diprotodontidae	1	0	0	1	1	1	1	1	1	1	1	1	1	0	0	1	1
Ilariidae	1	1	0	0	1	1	?	0	1	?	?	?	?	?	1	?	?
Vombatidae	1	1	1	0	1	1	1	1	1	1	1	1	1	1	1	1	1

	35	36	37	38	39	40	41	42	43	44	45	46	47	48	49	50	51
Microbiotheridae	0	0	1	0	1	0	0	0	0	0	0	0	0	0	0	0	0
Dasyuridae	0	1	0	0	1	0	0	0	0	1	0	1	0	0	0	0	0
Peramelidae	0	1	0	0	1	0	0	1	0	1	0	0	0	1	0	1	1
Pseudocheiridae	0	0	1	0	0	0	0	0	0	0	0	0	0	0	0	0	0
Phalangeridae	0	0	0	0	0	0	0	0	0	0	0	0	0	0	0	0	0
Phascolarctidae	0	0	1	0	0	0	0	0	0	0	0	0	1	0	0	0	1
Thylacoleonidae	0	0	1	1	0	0	0	0	0	0	0	1	0	0	1	0	1
Palorchestidae	0	0	1	1	0	0	1	1	0	0	0	1	0	1	1	1	1
Diprotodontidae	1	1	1	0	0	0	1	1	0	0	0	1	0	0	1	1	1
Ilariidae	?	1	1	0	1	1	1	1	1	1	1	0	1	?	?	?	1
Vombatidae	1	1	1	1	1	1	1	1	1	1	1	0	1	1	0	1	1

(Appendix III, cont.)

	52	53	54	55	56	57	58	59	60	61	62	63	64	65	66	67	68
Microbiotheridae	0	0	0	0	0	0	0	0	0	0	0	0	0	?	0	0	0
Dasyuridae	0	0	0	0	0	0	0	0	0	0	0	0	0	0	0	0	1
Peramelidae	1	1	1	0	0	0	1	1	0	0	0	0	0	0	0	0	1
Pseudocheiridae	0	0	0	0	0	1	0	0	0	0	0	0	0	0	0	0	1
Phalangeridae	0	0	0	0	0	0	0	0	0	0	0	0	0	0	0	0	1
Phascolarctidae	1	0	0	0	0	1	0	0	1	1	0	0	0	0	0	0	1
Thylacoleonidae	0	0	0	0	0	1	0	0	0	1	0	0	0	?	1	0	1
Palorchestidae	1	1	0	0	0	1	0	0	1	0	0	1	0	1	1	1	1
Diprotodontidae	1	1	1	1	1	1	1	0	1	1	1	1	1	1	1	1	1
Ilariidae	1	1	1	1	?	1	1	1	?	?	?	?	?	?	?	?	?
Vombatidae	1	1	1	1	1	1	1	1	0	1	1	1	1	0	1	1	1

	69	70	71	72	73	74	75	76	77	78	79	80	81	82	83	84	85
Microbiotheridae	0	1	0	0	0	0	0	0	0	0	0	0	0	0	0	0	0
Dasyuridae	0	1	1	1	1	1	0	0	0	0	0	1	0	1	0	0	0
Peramelidae	0	0	1	1	0	0	0	0	0	0	0	1	0	1	0	0	0
Pseudocheiridae	0	0	0	0	0	0	0	0	1	1	0	0	0	0	1	0	0
Phalangeridae	0	0	0	0	0	1	0	0	1	0	0	0	0	0	1	0	0
Phascolarctidae	1	1	0	1	0	0	0	0	1	0	0	1	1	0	1	0	0
Thylacoleonidae	1	1	1	1	1	1	0	0	1	0	0	1	1	0	1	1	1
Palorchestidae	1	1	1	1	1	1	1	0	1	1	1	1	1	0	1	1	1
Diprotodontidae	1	1	1	1	1	1	1	1	1	1	1	1	1	0	1	1	1
Ilariidae	?	?	?	?	?	?	?	?	?	?	?	?	?	?	?	?	?
Vombatidae	1	1	1	1	1	1	1	1	1	1	1	1	1	0	1	1	1

(Appendix III, cont.)

	86	87	88	89	90	91	92	93	94	95	96	97	98	99	100
Microbiotheridae	0	0	0	0	0	0	0	0	0	0	0	0	0	0	0
Dasyuridae	0	0	0	1	0	0	0	0	0	0	0	0	0	0	0
Peramelidae	0	0	0	1	0	1	?	0	0	0	1	0	0	0	0
Pseudocheiridae	0	0	0	0	0	0	0	1	0	0	1	0	0	0	0
Phalangeridae	0	0	0	0	0	0	0	1	0	0	1	0	0	0	0
Phascolarctidae	0	0	1	0	1	0	0	1	0	0	1	0	0	0	0
Thylacoleonidae	0	?	?	?	?	?	?	1	0	1	1	0	0	0	0
Palorchestidae	0	0	0	0	1	0	0	1	0	0	1	1	0	1	0
Diprotodontidae	1	1	1	1	1	0	1	1	1	0	1	0	1	1	0
Ilariidae	?	?	?	1	?	1	1	1	0	1	?	?	?	?	?
Vombatidae	1	1	1	1	1	1	1	1	1	1	1	1	0	1	1

Literature Cited

Aplin, K.P., and M. Archer.
1987   Recent advances in marsupial systematics, with a new syncretic classi-
            fication. *In* M. Archer (ed.), *Possums and Opossums: Studies in Evo-
            lution*. Roy. Zool. Soc. N.S.W., Sydney.

Archer, M.
1984a   The Australian marsupial radiation. *In* M. Archer and G. Clayton
            (eds.), *Vertebrate Zoogeography and Evolution in Australia*. Hesperian
            Press, Perth.

1984b   On the importance of being a koala. *In* M. Archer and G. Clayton
            (eds.), *Vertebrate Zoogeography and Evolution in Australia*. Hesperian
            Press, Perth.

_____ and A. Bartholomai
1978   Tertiary mammals of Australia: a synoptic review. *Alcheringa* 2:1-19.

Barbour, R.A.
1977   Anatomy of marsupials. *In* B. Stonehouse and D. Gilmore (eds.),
            The Biology of Marsupials. University Park Press, Baltimore.

Bartholomai, A.
1962   A new species of *Thylacoleo* and notes on some caudal vertebrae of
            Palorchestes azael. *Mem. Queensland Mus.* 14:2, 33-40.

Behrensmeyer, A.K., and D.E. Boaz
1980   The recent bones of Amboseli National Park, Kenya. *In* A.K. Beh-
            rensmeyer and A.P. Hill (eds.), *Fossils in the Making*. University of
            Chicago Press, Chicago.

Bensley, B.A.
1903   On the evolution of the Australian Marsupialia. *Trans. Linn. Soc.
            London* 9:83-217

Bryant, B.J.
1977   The development of the lymphatic and immunohematopoietic sys-
            tems. *In* D. Hunsaker II (ed.), *The Biology of Marsupials*. Academic
            Press, New York.

Case, J.A.
1989   Antarctica: the effect of high-latitude heterochroneity on the origin of
            the Australian marsupials. *In* J.A. Crame (ed.), *Origins and Evo-
            lution of the Antarctic Biota*. Geological Society special publ. 47:217-
            226.

Dublin, L.I.
1903   Adaptations to aquatic, fossorial and cursorial habits in mammals. II.
            Arboreal adaptations. *Amer. Naturalist* 37:443, 731-736.

Elftman, H.O.
1929   Functional adaptations of the pelvis in marsupials. *Bull. Amer. Mus.
            Nat. Hist.* 58.

Felsinstein, J.
1987   *PHYLIP Manual*, Version 2.8. University Herbarium, University of
            California, Berkeley.

Finch, M.E.
1971   *Thylacoleo*, marsupial lion or marsupial sloth? *Austral. Nat. Hist.*
            17:1, 7-11.

1982   The discovery and interpretation of *Thylacoleo carnifex* (Thylaco-
            leonidae, Marsupialia). *In* M. Archer (ed.), *Carnivorous Marsupials*.
            Roy. Zool. Soc. N.S.W., Sydney.

_____ and L. Freedman,
1986   Functional morphology of the vertebral column of *Thylacoleo carnifex* Owen (Thylacoleonidae, Marsupialia). *Austral. J. Zool.* 34:1-16.

Flannery, T.
1983   A unique trunked giant. *In* S. Quirk and M. Archer (eds.), *Prehistoric Animals of Australia*. Australian Museum, Sydney.

Forbes, W.A.
1881   On some points in the anatomy of the koala. *Proc. Zool. Soc. London*, pp.180-195.

Gray, H.
1977   *Gray's Anatomy*, T.P. Pick and R. Howden, eds. Bounty Books, New York, 1977 ed.

Gregory, W.K.
1910   The orders of mammals. *Bull. Amer. Mus. Nat. Hist.* 27.

1951   *Evolution Emerging*, vol. 1. MacMillan, New York

Harding, H.R., F.N. Carrick, and C.D. Shory
1979   The affinities of the koala *Phascolarctos cinereus* (Marsupialia, Phascolarctidae) on the basis of sperm ultrastructure and development. *In* M. Archer (ed.), *Possums and Opossums: Studies in Evolution*. Roy. Zool. Soc. N.S.W., Sydney.

Hildebrand, M.
1988   *Analysis of Vertebrate Structure*. John Wiley & Sons, New York.

Hughes, R.L.
1965   Comparative morphology of spermatozoa from five marsupial families. *Austral. J. Zool.* 13:533-543.

Huxley, T.H.
1880   On the application of the laws of evolution to the arrangement of the Vertebrata, and more particularly of the Mammalia. *Proc. Zool. Soc. London*, pp.649-662.

Kirsch, J.A.W.
1977   The comparative serology of Marsupialia, and a classification of mar-
                supials. *Austral. J. Zool.* suppl. series no. 52:1-152.

Lewis, O.J.
1964   The homologies of the mammalian tarsal bones. *J. Anat.* 98:195-208.

1980a   The joints of the evolving foot, part 1. The ankle joint. *J. Anat.*
                130:527-543.

1980b   The joints of the evolving foot, part 2. The intrinsic joints. *J. Anat.*
                130:833-857.

1983   The evolutionary emergence and refinement of the mammalian pat-
                tern of foot architecture. *J. Anat.* 137:21-46.

Marshall, L.G., J.A. Case, and M.O. Woodburne
1990   Phylogenetic relationships of the families of the Marsupials. *In* H.H.
                Genoways (ed.), *Current Mammalogy*, vol. 2. Plenum Press, New York

Owen, R.
1838   On the osteology of the Marsupialia. *Proc. Zool. Soc. London* 6:120-
                149.

1874   On the osteology of the Marsupialia. *Trans. Zool. Soc. London* 8:483-
                500.

1883a   On the fossil mammals of Australia, part 15. On the affinities of
                Thylacoleo. Philos. Trans. Roy. Soc. London, pp.557-582.

1883b   On the fossil mammals of Australia, part 19. Pelvic characters of
                Thylacoleo carnifex. Philos. Trans. Roy. Soc. London, pp.639-643.

Pledge, N.S.
1987   *Muramura williamsi*, a new genus and species of ?Wynyardiid (Marsu-
                pialia, Vombatoidea) from the middle Miocene Etadunna formation
                of South Australia. *In* M. Archer (ed.), *Possums and Opossums:
                Studies in Evolution.* Roy. Zool. Soc. N.S.W., Sydney.

Rauscher, B.
1987   *Priscileo pitikantus*, a new genus and species of Thylacoleonid marsu-
               pial (Marsupialia, Thylacoleonidae) from the Miocene Etadunna for-
               mation, South Australia. *In* M. Archer (ed.), *Possums and Opossums:
               Studies in Evolution.* Roy. Zool. Soc. N.S.W., Sydney.

Rich, P.V., G.F. Van Tets, and F. Knight
1985   *Kadimakara: Extinct Vertebrates of Australia.* Pioneer Design Studio,
               Victoria.

Ride, W.D.L.
1964   A review of Australian fossil marsupials. *J. Proc. Roy. Soc. W. Austral.*
               47:97-131.

Shimer, H.W.
1903   Adaptations to aquatic, arboreal, fossorial and cursorial habits in
               mammals. III. Fossorial adaptations. *Amer. Naturalist* 37:44.

Sisson, S.
1914   *The Anatomy of the Domestic Animals.* W.B. Saunders Co., Phila-
               delphia.

Sonntag, C.F.
1923   On the myology and classification of the wombat, koala and pha-
               langer. *Proc. Zool. Soc. London* for 1922, pp.863-869.

Spence, A. and E. Mason
1983   *Human Anatomy and Physiology.* 2nd ed. Benjamin/Cummings Publ.
               Co., Menlo Park, Calif.

Springer, M.S.
1988   The phylogeny of diprotodontian marsupials based on single-copy
               nuclear DNA-DNA hybridization and craniodental anatomy. Ph.D.
               thesis, University of California, Riverside.

Stirling, E.C.
1913   Fossil remains at Lake Callabonna, part 2. On the identity of *Phas-
               colomys (Phascolonus) gigas* Owen and *Sceparnodon ramsayi* Owen,
               with a description of some remains. *Mem. Roy. Soc. S. Austral.* 1:111-
               178.

_____ and A.H.C. Zietz,
1899   Fossil remains at Lake Callabonna, part 1. Description of the manus
             and pes of *Diprotodon australis*, Owen. *Mem. Roy. Soc. S. Austral.* 1:1-
             40.

Stirton, R.A.
1967   The Diprotodontidae from the Ngapakaldi Fauna, South Australia.
             Austral. Bur. Miner. Resour., Geol. and Geophys. Bull. no. 85.

Strahan, R.
1978   What is a koala? *In* T.J. Bergin (ed.), *The Koala: Proceedings of the
             Taronga Symposium on Koala Biology, Management, and Medicine,*
             Sydney: Zoological Parks Board of N.S.W., pp.3-19.

Szalay, F.S.
1977   Ancestors, descendants, sister-groups and testing of phylogenetic
             hypotheses. *Syst. Zool.* 26:12-18.

1981   Functional analysis and the practice of the phylogenetic method, as
             reflected by some mammalian studies. *Amer. Zool.* 21:37-45.

1982   A new appraisal of marsupial phylogeny and classification. *In* M.
             Archer (ed.), *Carnivorous Marsupials*, pp.621-640. Roy. Zool. Soc.
             N.S.W., Sydney.

1984   Arboreality: Is it homologous in metatherian and eutherian mam-
             mals? *Evol. Biol.* 18:215-258.

Tedford, R.H., M. Archer, A. Bartholomai, M. Plane, N.S. Pledge, T. Rich, P. Rich,
and R.T. Wells,
1977   The discovery of Miocene vertebrates, Lake Frome area, South Aus-
             tralia. *Austral. Bur. Miner. Resour., Geol. Geophys. Journal.* 2:53-57.

Tedford, R.H., and M.O. Woodburne
1987   The Ilariidae, a new family of vombatiform marsupials from Miocene
             strata of South Australia and an evaluation of the homology of molar
             cusps in the Diprotodonta. *In* M. Archer (ed.), *Possums and
             Opossums: Studies in Evolution.* Roy. Zool. Soc. N.S.W., Sydney.

Waters, B.T.
1967   Osteology of *Diprotodon*. M.A. thesis, University of California, Berkeley.

Wells, R.T., and B. Nichols
1977   On the manus and pes of *Thylacoleo carnifex* Owen (Marsupialia). *Trans. Roy. Soc. S. Austral.* 101:6, 139-146.

Woodburne, M.O.
1984   Families of marsupials: relationships, evolution and biogeography. *In* T.W. Broadhead (ed.), *Mammals: Notes for a Short Course*. University of Tennessee Dept. Geological Sci., Studies in Geology no. 8, pp.48-71.

_____, R.H. Tedford, M. Archer, W.D. Turnbull, M.D. Plane and E.L. Lundelius
1985   Biochronology of the continental mammal record of Australia and New Guinea. *Spec. Publ. S. Austral. Dept. Mines and Energy* 5:347-363.

DATE DUE

DEMCO 38-297